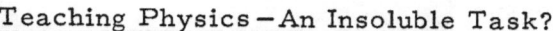
Teaching Physics—An Insoluble Task?

Teaching Physics—An Insoluble Task?

Proceedings of the International Congress
on the Education of Teachers of Physics in
Secondary Schools, Eger, Hungary, September 1970.

Edited by
Sanborn C. Brown
F. J. Kedves
E. J. Wenham

The MIT Press   Cambridge, Massachusetts, and London, England

MIT Press

0262020785

BROWN
TEACH PHYSICS INSOL

Copyright © 1971 by
The Massachusetts Institute of Technology

This book was designed by The MIT Press Design Department.
It was printed and bound in the United States of America.

ISBN 0 262 02 0785 (hardcover)

Library of Congress catalog card number: 71-148856

Contents

Every modern scientist, to a greater or lesser degree worries about the future of his subject, wonders about the effect of what he is doing on the people about him, and what in the long run science and its attendant technology are doing to society. Teachers are perhaps more concerned about these aspects of their profession than most, because they are in direct contact with the youth of the world which is that segment of our society that not only will have to guide it in the future, but which today is bringing into sharp focus its dissatisfaction with what appears to be the result of our ever increasing acceleration of scientific progress.

Basically these are not national but international problems, and the underlying theme of any gathering of science teachers from anywhere in the world is how to understand problems of this nature, and how to come to terms with the interaction of science and the needs and aspirations of their students.

In September of 1970 (the 11th to the 17th) an International Congress on the Education of Teachers of Physics in Secondary Schools met in Eger, Hungary, for the specific purpose of reviewing the problems of recruitment and educating prospective teachers of physics in secondary schools and of assisting teachers to keep up to date, to consider solutions that have been arrived at in certain countries, and to assist those who are trying to solve the problems in their own local situations. Inevitably the discussions of specifics were carried out against the background of concern for the place of science in general, and physics in particular, in the rapidly changing climate of modern society. This whole tone of the Conference was set by the opening address of Professor P. L. Kapitza, member of the Academy of Sciences of the U.S.S.R., and followed up throughout the week by a combination of invited papers, background material contributed by members of the Congress, and working groups of concerned participants. These working groups not only met daily on specific matters of their own personal interest, but developed guidelines for suggested action which were presented to and discussed by the whole Conference.

This report of the Congress's deliberations takes the following form. Because the larger issues of the relevance of the teaching of physics to the problems of society underlay all the discussions, the basic ideas which were presented to the Conference on these subjects are given in the first chapter so that, against this background, specific areas of concern on the education of teachers may be viewed in their proper perspectives. Subsequent chapters begin with the formally recommended guidelines which resulted from the deliberations of each "working group." Then are presented the statements and discussions that led to the formulation of the guidelines.

Although the editors have tried hard not to change either the meaning or the style of the material as it was given, to produce some literary uniformity, in all cases the written material has undergone some editing, and the spoken material, which was transcribed from recordings made at the time of delivery, was polished and revised to render it more acceptable as a written record.

This was basically, however, a conversational conference. It was in the give-and-take of discussions that the most stimulating ideas were expressed, and the informality which invited free flow of information was further enhanced by the invited speakers, many of whom spoke only from notes or from prepared texts in languages foreign to those in which they talked. The editors have attempted to maintain this spirit in these proceedings. In presenting the material, extensive polishing of English was purposely avoided since we wanted to retain the conversational tone of the Conference as it was actually conducted.

All those who have been to meetings of this sort know very well that extemporaneous comments from individuals, thinking hard about the subject matter at hand, lack a certain free flow of context from one speaker to the next. Nevertheless, it is just this kind of talk from many different countries and many different kinds of educational systems that greatly enriched the value of the whole conference. Because of this, in taking material off of the recording tape, although the editors have grouped subject matter together, they have preserved the somewhat disjointed nature of the interchanges between conferees in an attempt to recapture some of the stimulation of the face-to-face discussion of many different ideas and ways of attacking the common problems of the education of teachers.

The wide representation at Eger by delegations from all over the world is well illustrated by perusing the names and addresses of those who attended the Conference. These are given in Appendix B. A brief explanation of why this same distribution does not appear among the invited and contributed papers is perhaps in order. The Program Committee invited the list of speakers, not primarily to balance international representation, but to bring to the conference the stimulating focus on the subject matter which it so successfully achieved. The contributed papers were entirely in the hands of those who volunteered to write them, and it is apparent that both the knowledge of the existence of the Conference, and the time and desire to prepare a contribution varied widely from country to country. However, the form which the Conference took, where major emphasis was placed on the working groups, makes the identification of those who made some very major contributions to the thinking of the conferees less evident in these published proceedings than would have been the case in a mere compilation of submitted papers.

The working groups had excellent international representation, and the impact of these reports lay specifically in their value as coming from the whole group. It is quite impossible, and in fact it would have been undesirable, to identify individual contributions. Not only did the working groups produce finished reports, but as part of their deliberations they greatly aided the work of the editors of these Proceedings by identifying by specific reference those part of the invited and contributed papers which gave particular emphasis to detailed points in their group reports. As a result, the editors did not have to rely solely on their own judgment as to what was germane to the issues discussed but could reflect the cooperative wisdom of the whole Conference.

The labor and cooperation of a large number of people went into producing these Proceedings. For the week following the end of the Conference, the editors stayed in Eger to be close to the facilities of the Conference Center. We are particularly grateful to the Center staff for all of the help provided, and most especially to Mr. and Mrs. István Hriszó. When we were in need of surveying a large number of tape records of speeches in Russian, three students in their final year at the Ho Chi Minh Pedagogical Institute of Eger, Miss Dalma Fülöp, Miss Ágnes Dezső, and Miss Katalin Kovács, cheerfully spent many hours translating these into Hungarian which helped us immensely and for which we are very grateful. Further help in translating was also given to us by Miss Erzsébet Fritz, for which we wish to thank her.

After our second week in Eger, it became evident that the
need for additional secretarial help and translation facilities
could be achieved more easily by moving closer to the Eötvös
University in Budapest, and this change was made. Little did we
realize, however, that the hotel facilities of Budapest were far
overtaxed, and we were unable to find hotel accommodations. As
a result, we invaded en masse the quiet home of the Kedves family, for whose hospitality to typewriters, tape recorders, secre-
taries, and editors, and all the resulting noise and confusion we
owe a lasting debt of gratitude.

Our constant helper, secretary, and translator was Miss
Rita Ignáczy, whose patience, good humor, and ability not only
made it possible to finish the Proceedings while the editors were
still in Hungary, but made more enjoyable our long hours and
concentrated labor. We hope that her total immersion in English
for the weeks she was with us will be useful to her in her chosen
future as a language teacher. We also want to thank Mrs. Istvan
Pataki for her skillful typing of a large part of the final draft even
though she did not know English.

Finally, it is with a deep sense of gratitude that we acknowl-
edge the great contribution of Lois Brown (Mrs. S.C.B.) who
was again willing to give up any thought of a relaxing holiday to
be the ever-willing checker, language polisher, auxiliary typist,
and the general expediter of an international conference proceedings.

S.C.B.
F.J.K.
E.J.W.

Budapest, Hungary
October 1970

Congress Organization

Program Committee

D. Sette (Italy) Chairman
R. B. da Costa (Brazil)
A. Harashima (Japan)
R. L. Krans (Netherlands)
E. Nagy (Hungary)
J. A. O. Sofolahan (Nigeria)
F. G. Watson (U.S.A.)

Organizing Committee

E. Nagy (Hungary) Chairman
B. Fogarassy (Hungary)
F. J. Kedves (Hungary)
T. Mátrai (Hungary)
E. Sas (Hungary)

Exhibits Committee

T. Mátrai (Hungary) Chairman
J. Boutigny (France)
H. Jensen (U.S.A.)

Funds for the Congress were provided by the Hungarian Government, the International Union of Pure and Applied Physics, (IUPAP), and UNESCO. The Congress was sponsored by the International Commission on Physics Education of IUPAP.

September 11, Friday

3.30 P.M.
Opening addresses:
E. Nagy,
Chairman of the Organizing Committee
H. H. Staub,
President of the Commission on Physics Education of the International Union of Pure and Applied Physics
L. Tamás,
Secretary of the Hungarian Socialist Labor Party of Heves County
A. Kónya,
Hungarian Academy of Sciences
I. Kovács,
Eötvös Loránd Physical Society
J. Fekete,
Hungarian Ministry of Education
T. Mátrai,
Chairman of the Exhibits Committee

4.00 P.M.
P. L. Kapitza (U.S.S.R.):
"General Principles of the Education of Present-day Youth and General Methods of Secondary-School Physics Teaching"

5.00 P.M.
D. Sette (Italy):
Organization of Working Groups

September 12, Saturday

Session Chairman:
J. A. O. Sofolahan (Nigeria)

8.30 A.M.
W. Kroebel (Federal Republic of Germany):
"The Training of Physics Teachers for Secondary Schools and the Dependence of this Training on the Instruction in Universities"

9.30 A.M.
N. Joel (UNESCO):
"How Can International Organizations Help Physics Teachers"

11.00 A.M.
E. M. Rogers (U.K.):
"The Use of New Examinations and of Examination-Construction
Seminars in Curriculum Revision and in the Training of Teach-
ers"

September 14, Monday
Session Chairman:
R. B. da Costa (Brazil)

8.30 A.M.
R. N. Little (U.S.A.):
"Pre-Service Formation of Physics Teachers: Technical Educa-
tion"

9.30 A.M.
F. Watson (U.S.A.):
"Pre-Service Pedagogical Formation of Physics Teachers"

11.00 A.M.
S. C. Brown (U.S.A.):
J. Zemplén (Hungary): Report on the International Working Semi-
nar on the Role of History of Physics in Physics Education,
Cambridge, Massachusetts, July 1970

11.30 A.M.
Working Group Reports

2.30-5.30 P.M.
Working Groups

September 15, Tuesday
Session Chairman:
H. H. Staub (Switzerland)

8.30 A.M.
E. J. Wenham (U.K.):
"The In-Service Education of Physics Teachers"

9.30 A.M.
Panel Discussion:
"How can secondary school teaching be made more responsive
to the new needs of modern society?"
M. Y. Bernard (France), S. G. Bronevshuk (U.S.S.R.), A.
Harashima (Japan), E. Nagy (Hungary), R. S. Tilton (U.S.A.),
S. C. Brown (U.S.A.), Moderator

11.00 A.M.
Working Group Reports

2.30-5.30 P.M.
Working Groups

September 16, Wednesday
Session Chairman:
J. Fuka (Czechoslovakia)

8.30 A.M.
M. W. P. Strandberg (U.S.A.):
"Technology in Education"

9.30 A.M.
G. Marx (Hungary):
"An Insoluble Task: Teaching Physics"

11.00 A.M.
Working Group Reports

2.30-5.30 P.M.
Working Groups

September 17, Thursday
Session Chairman:
W. C. Kelly (U.S.A.)

8.30 A. M.
Final Working Groups Reports

11.50 A. M.
Adjournment

Teaching Physics—An Insoluble Task?

Chapter 1

The Response to Modern Society

The Conference was fortunate in having as its opening speaker,
Professor P. L. Kapitza of the Soviet Union. He set the tone of
informality for the whole conference by giving his paper in
English, using a Russian text only as notes. His subject was
"The General Principles of the Education of Present-day Youth
and General Methods of Secondary-School Physics Teaching":

I accepted the invitation to speak to this conference with
some reluctance because I am not a teacher and have never
taught any secondary-school teachers. But I take great interest
in how young people are taught and therefore I have come to
speak on general questions of education.
Before talking of teaching and education, you must first of
all take account of the social changes which have been happening
during these last years. Science and technology have always had
great influence on culture. But during the recent years, the
most recent twenty to thirty years, this influence has been so
great that the scientific achievement has influenced the social
structure of the world on a large scale. This influence is now
called the scientific-technological revolution. Now, more and
more, social phenomena can be regarded as part of the tech-
nical-social revolt. This has also had a great influence on education, and two particular effects on physics.
You know very well that the effect of science on social conditions has been the tremendous increase in productive power of the
human being. This was mainly due to the fact that manual power
was replaced by motors and by big sources of electrical power.
When you add automation, you will see that this productive
power increased immensely in industry as well as in farming
and agriculture. For instance, during the last century 80 to 90
percent of the population was engaged in farming, in producing
food; only 10 percent lived in towns. Now only 10 percent of the
population in America is engaged in producing food and the
others can be engaged in industry and production, and the productive power per person is high. For instance, if you consider
a modern automobile factory and divide the number of automo-

biles it produces by the number of all workers in the plant, you
will find that the work of one person produces more than one
automobile per month.

Our modern economists reckon that only a quarter of the
present population of a country is needed in industrial production to supply all of the population with food, clothing, housing,
and necessary services. A number of the rest of the population
may be engaged in war industry, in helping less developed
countries, and in activities like sport, cinema, television, and
traveling.

Such a rise in productivity means that we can produce more
than we need to live. This is all very interesting and has influenced education very much. In the last century in England, at
that time the most industrially developed country, only a very
small number of well-to-do people could go to a university and
be educated to the age of 20 or 23. Faraday became an appren-
tice in a book-printing establishment when he was 14. So did
most other people go to work early. Education was limited for
the general population to 14 years, and the work day very often
grew to be 12 or 14 hours long. Now it has become economically
possible to educate all the population not only through secondary
school up to the age of 17 or 18, but to give them a higher education. Also the number of students in the universities is growing
very rapidly. In the United States, in the Soviet Union, and in
all the developed countries, the number of students in the universities has doubled in the past 10 years. The numbers are
changing all the time but are always growing, and if you extrapolate at this moment you will find that at the end of this century
you can expect that all the people of these countries may have
higher education. Such a possibility of universal higher education has a big influence on the secondary school, which would
come to be regarded only as a preparatory stage for the higher
school.

There is another effect of the scientific-technological revolution on education which is much more subtle. During the last
century the leading economists reckoned that the increase of
production would obviously be accompanied by the impoverish-
ment of the proletariat. However, the scientific-technological
revolution changed this prediction. The increase of productivity
per person was so great that there was a tremendous increase

of wealth, of income per person in all countries. Unemployment
and poverty in some countries is due only to the social structure,
but I do not mean that with the present economy's means of production this could be avoided easily. But with this high productivity and high income for most people there arises another problem, the problem of leisure. A very important problem.

To illustrate, you can put this problem in the following way.
If at present a person is engaged in his work about 6, 7, or 10
hours a day, and if he sleeps a normal amount of time and uses
only about two hours for eating and moving from place to place,
he has 7 or 8 hours left for living. What can a man do in this
time? Has he ways to spend his leisure reasonably and properly?
This is a very important question and a great social problem.
Especially because this time of leisure will eventually grow
longer and longer. A number of economists predict that the introduction of the computer and all the electronic devices will
shorten the working time and increase the leisure even more,
and the time of leisure will be longer than the time of work.

It is interesting that the first to tackle the problem of leisure was Aldous Huxley in his very interesting book, Brave New
World. This is a utopia, and if you have read this book you will
remember that he solves the problem in the following way: The
general population must use their leisure for sporting entertain-
ment, for all sorts of very elemental shows and for sex. And
finally, if they are not easy enough to control a great emphasis
is put on narcotics. The government of the "brave new world"
was very careful to educate all the working population so that
they had no spiritual demands, no spiritual concern for social
conditions. For this purpose they were taught by use of the
Pavlov reflex and the Freudian subconscious to despise science
and culture.

These predictions of Huxley seem to be proving accurate in
the most developed countries, like the U.S.A. The change is not
in the decline of culture, not in the decline of civilization, but in
the rise of crime and in the rise of the use of narcotics. This
leisure is most badly used by the very young population. When
they reach maturity, some of them—not all—are mostly interested in sports and some very elementary shows and entertain-
ment. There is no barrier for sex and there is easy access to
all sorts of gadgets like motorcars, cameras, cinema, and such

things which are used in such a very primitive way that interest
in them is very quickly saturated.

These young people think that their parents are well-to-do
and that they will never be put in a position of poverty. They
have no need to think of their own futures, to think how to increase their own control of what they can achieve through work.
The capitalisitic society develops in them a selfish individualism. They have no interest in social problems. So they turn to
the use of narcotics, which give only temporary relief, poison
the nervous system, and actually lead gradually but deliberately
to the growth of criminals. It is quite natural that the best part
of the young people beginning to protest, try to be hippies; you
can see the first symptom of such protest in the hippies and the
beatniks, with whom you are familiar. But indeed this is not a
serious movement, with social consequence. However it is clear
that such phenomena as hippies and beatniks may appear only in
an affluent society, because when everybody has to work hard
such things cannot happen.

Much more serious and much more important is the student
unrest which is appearing now in all countries. This has great
political significance. For example, in the United States data
show that 55 percent of the schoolboys who finished secondary
school last year went to the university, and we now have in the
U.S.A. about 5 million undergraduates. This is a considerable
political force. It is interesting that the study of student unrest
in the U.S.A. shows that the people who take the more active
part in this movement come from well-to-do families. Thus this
movement cannot be regarded as a protest about economics. It
is purely an ideological one. A purely social one. People are
simply not satisfied with the state of social conditions, the ideology with which they are surrounded. They have no sense of
direction as to where to aim their activities and their energies,
these young people. And neither the country nor the social system provides it.

At present there is no definite ideology in this movement.
It is purely a general discontent among the students. The ideology comes later. If you study all the revolutions, you will find
that the discontent comes first, and then the aims. The students
do not yet have the aims, but these are to come. This all shows
that our present social structure is not prepared for so much

wealth along with so much leisure. The social structure must
change. This general leisure brings its own social changes. The
question of leisure is being studied by a number of sociologists,
social workers, and other investigators. A number of Americans
feel there is no way out. There is no way to stop the growth of
productivity, the industrial growth of all countries, and they see
in this movement the end of present civilization. They all agree
that the question of leisure is as important as the question of
peace in atomic war, for the survival of humanity.

I think we have no definite reason to be so pessimistic.
There are two ways out of the present position. The first is to
develop further the idea of Huxley, giving only a part of the
people a reasonable education and keeping the others in a semi-
animal state. The animal desires will, in a crude and primitive
way, stop them from having a reasonable thinking and cultural
life. The other direction is the opposite, to educate all the people
to such an extent that they can choose reasonable and useful interests for their leisure. This brings us to the matter of education. Only by means of education can we solve it provided every
developed country will indeed choose this way to oppose the wrong
social condition now produced by this continuing increase of leisure and wealth resulting from the scientific-technical revolution.

It is clear that we must change education in order to educate
people properly in ways to use their leisure. Up to now, we have
always approached education pragmatically. We teach a young
person to be a good doctor, a good lawyer, a good engineer, a
good designer. We teach them for the practical world. Now we
must start to teach people to use their leisure. It is much more
important to teach people to use their leisure than to do their
work. This is as true in secondary education as in higher education. What sort of education must it be, to make a man cultured?

This is a new problem and I cannot deal with its details, but
I think I can describe the general direction in which this education must go. I think you have all experienced the idea that the
happiest people in America are those who do creative work: scientists, writers, painters, artists, film producers. You know
that people working in art who want to do creative work do not
divide their time between leisure and work. All is equally interesting and they do not know which is leisure and which is work.

Therefore to make a man happy during his leisure, you must teach him to do creative work. Now we must define the word "creative" in a much broader way than is usual. "Creative" means that in any work in which he is engaged a man should be able to find a solution for himself and within himself. If he acts according to a definite instruction from someone else, he is not creative. Present mass production is arranged in such a way that a man cannot do any creative work on an assembly line, but has to work by quite definite instruction, and any small change will produce a negative effect. So the work of mass production is dull and tedious for the worker. If you remember the masterful movie of Charlie Chaplin, Modern Times, you will know what is really meant by mass production.

At the same time, our experience of life shows that it is quite possible to spend our leisure in interesting ways if we do creative work. Only a few people do it now. We have the natural ability to be creative. I know a number of people who spend their leisure very reasonably as artists or painters, but who find creative work also in their environment, in social problems, in traveling. What education must teach the general bulk of people is how to do creative work by themselves and enjoy it.

I should explain that while you are doing creative work with proper education, you will enjoy it. Let us say, you spend your leisure in traveling. A number of people do it. It is very common to travel all around the world, and you come to some town, ancient or modern. You look at it. If you look at it knowing the history of this town, knowing the history of the people who live in this town, you can make judgments very different from what you would think if you did not know all this. It is much more interesting to you. But even further, you can compare the ancient town with the modern, the modern way of life with the old. This is even more interesting. If you are taught how to compare and if it is your natural interest, it is more interesting still.

Another new problem put before educators is how to give a man a large amount of knowledge and ability to develop independent thinking about it, to be able to comprehend the environment in a creative way. The creative ability of a man appears in very early days, in the school, but the period when the most fruitful interests become clear is probably after the age of eight. So the secondary school must provide a general development of the cre-

ative ability of a man, and the university must find the region in
which it is to be best applied. The secondary school must help a
young man develop creative ability in any subject, physics, mathematics, and all.

This question of developing creative ability in young people
interested me long before this problem of leisure appeared. I
was interested mainly as a scientist who wanted to have good
students and develop science. All you need for your pupils and
for your research are people who have been developing creative
ability for a long time.

Now we must see what changes must take place in the secondary school if at present most young people are taught to
memorize a definite number of facts. The student must be led
toward independent thinking, toward creative activity. This kind
of education requires you to approach each person individually,
not generally. We need an independent teaching system that is
much more complicated and much more difficult than the present
system. You will soon know which student is interested in natural
science and which in art. The school must separate the students
and take better advantage of the differences; it must teach the
pupil himself.

To focus on the pupils who are to be taught natural science,
we see that the teaching of creative ability and independent thinking has three aspects. First you must teach the young people how
to generalize phenomena, the method of induction. Second, you
must teach how to predict the phenomena of nature from theoretical generalizations, the method of deduction. Finally, you
must teach the student how to look at the contradictions in nature
and to solve these contradictions, the method of dialectics. The
best way to teach these aspects is by means of solving problems
and working in laboratories. So we can see at once that the best
subject for developing creative thinking in a young man is physics,
and this is what makes the role of the teacher-physicist so important in their education.

It is very important to have laboratories and seminars and
to solve problems which will encourage independent thinking.
The exercises given to schoolboys are not always good for this.
Mostly such problems give the pupil the data and he is to put the
information into the proper formula to produce the correct answer. The work is to find the correct formula. This is not really
independent thinking.

I used to give my students at the university a different kind of problem.[1] I will simplify a few examples for secondary-school levels. (1) The power of a motor necessary to move a pump will give a jet of water large enough to extinguish a fire in a six-story building. With this problem the pupil himself must judge the height of the building and the size of the jet, how far it must reach to the building, and where to put the pump. Each pupil may decide differently on these points, and it is easy to see which is better. (2) What is the size of a convex lens which will focus sunlight in a spot where you will make a furnace. The pupil must find out what heat is necessary and how good the focus must be in determining the size of the lens, and once again he will be thinking out the whole problem. There are other less obvious problems which interested me and on which I would like to advise you.

There has been much interest in the Soviet Union and in other countries in developing schools for the most gifted young people. We collected the most talented in physics, biology, and mathematics and tried to teach them together in order to produce much better scientists. I have found that this is an absolutely wrong idea, often difficult, and never good. If you take an able school-boy from a school, you at once lower the level of teaching because he is the first assistant to the teacher in that he helps his schoolmates. More than this, when the boy teaches his friends, he teaches himself. The best way to learn something is to explain it to someone else. In a special school the able student loses this opportunity and thus he himself does not develop as quickly as in an ordinary school. Since boys in such a school also soon become conceited, things are even worse.

The second point to consider is the teacher. To teach creative physics in a secondary school, the quality of the teacher must be very high, and it is difficult to find enough good teachers. This problem is still possible to solve if its gravity is recognized. We have to see that in teaching school it is not enough to

----

1. Typical problems of this sort have been published by Kapitza; for example, see Nauka i zhizny 33, 1967. The problems are stated in the January issue, p. 122. The solutions are given in February, p. 156, March, p. 144, April, p. 140, May, p. 125, and June, p. 130.

educate the students, the undergraduates, but we must teach the teacher as well. The teacher need not be a professional teacher. He must be a scientist himself and then he will grow together with the students.

With these ideas, we organized a university and a high school in Moscow in which all these undergraduates got the general courses over the first two years, but then went to different institutes where they joined different laboratories and were taught with the research students themselves. Each research worker was given two or three students whom he taught, and for this he received a good salary. All this took more than one day a week, but it proved a great success. This institute produced a great number of academicians, and this program now exists in all our universities.

In the history of science it is curious that most of the great discoveries were made while the scientist was teaching people. Mendeleev found the periodic system while he was trying to arrange the elements in a way which undergraduates could understand easily. Lobashevsky in mathematics gave a course in elementary geometry to adults and was trying to explain to his own satisfaction the a priori evidence of the postulate of two parallel lines when he discovered non-Euclidean geometry. Stokes never proved his famous theory but gave it to his undergraduates in a collection of examples. When we refer to Stokes we refer to this collection he made for students. In a more modern illustration, I have a story told me by de Broglie himself. He was asked to explain his work to some research students at the Technical University in Zürich. When he tried to explain it in a reasonably simple way to undergraduates, he discovered his famous fundamental formulation.

You can see how it would be if the teachers in all the secondary schools were young research students. It is difficult to organize education in this way, because you have so many students. A pupil-research student contact may last more than one day at a time, but this cannot disturb the research work and the teacher will grow with his pupils. Creative ability in physics must be educated in the final stages not by professional teachers but by some device such as ours, which is not easy to arrange but which I think is very important.

I think what I have told you may be helpful in your work, but remember that all these things can begin only when you recognize the importance of educating creative ability in the young people. I would like to underline the idea that to teach today's young people you must approach them according to their abilities to develop their own creativities, and this must be done in the secondary schools as well as in the primary schools. This is really not just a small problem. This is a big problem, on which depends the future of our civilization. Our future generally depends upon the education of young people, but in the present state of society, when they have access to wealth and leisure, the people must learn to use them properly. Without this, we will have a catastrophe like an atomic war. All of us, scientists, teachers, and people in general, must approach this task very seriously and regard it as one of the most important problems of our civilization.

Kapitza's paper stimulated a great deal of discussion throughout the whole Conference on how secondary-school teaching could be made responsive to the new needs of modern society. Statements on the subject were made on a number of occasions, sometimes as part of a panel discussion which was arranged specifically on the subject, and sometimes as part of general discussions after some of the invited papers. To illustrate the wide range of statements on this subject, we have abstracted a number of these as useful contributions toward finding solutions to this all important problem. B. R. Chapman (U.K.) said:

I think we have during the last one hundred years concentrated in our curriculum reform on the content of science and on the process of science. Sometimes the balance has been toward content, at other times toward the process of science. In times of great technological change it seems to me that we push the content. At times when we have problems with our supply of scientists, we say: "Ah, it's because we have given too much content, let's give them process and that will change the situation." And we oscillate between the two. There is also a third element which seems to me to be vital importance to contemporary society, and this is the interaction between science and society, the way science influences society, the way society influences science, and the whole interaction between the two. The complex

decisions that society has to make when technological changes
are involved, the problems you have when you decide to build
the Concorde, the problems you have in society when you decide
whether or not to join in with the rest of Europe to accelerate a
program of one sort or another, these are the sort of problems
which have aspects that can be termed sociological and aspects
that can be termed scientific.

But we are in danger of letting our children leave school
with a simplistic idea of what science is. We ought to let them
see that the decisions scientists make, decisions that economists
make, which are affecting them both at present and when they
grow up, are related to science and that they do not in fact have
clear-cut answers.

M. Y. Bernard (France) felt that the decline of interest in
physics in favor of such subjects as social science, sociology,
psychology, and so on might well be due to the very mathematical
nature of their physics, and that physicists were not trying hard
enough to counteract this tendency. But, more generally, he
pointed out that the young people were so accustomed to airplanes,
television, the telephone, and so forth that they were not the least
interested in them. On the other hand the modern mass media
brought home to them that everything was not right with the world
and that the real problems of the world were the struggle against
pollution, against poverty, and the social relations between people
and that these things appeared to be made worse by technology.
He felt that if we did not make physics more interesting and more
exciting we would find ourselves in the position of the professors
of Latin and Greek who have no pupils because none of the children
are interested in these subjects. Professor Bernard went on to
propose that it was essential that all students be taught the basic
elements of physics, not only those who were enthusiastic but
those who were poor at it—even those who were bored, because
in the modern world you could never tell who was to become the
director of a company, a minister of state, a social worker, or
even a revolutionary. He felt that it was essential for everybody
to know that the laws of physics are rigid, that science obeys
strict laws, and that one cannot find these truths by just reading
philosophy.

A. Harashima (Japan), after pleading for creativity and flexibility, since the world was changing so fast that you could not

tell what particular areas of physics might be the most relevant
at any particular time in the future, said that he felt that through
the medium of television one could keep both teachers and students up to date with the rapidly changing priorities of society.

   In amplifying this idea, I. Bukovszky (Hungary) went on to
say:

   Students must have their say, but that is not enough; I think
that the general public must be given the opportunity to take an
interest and have their views known about school work and about
the process of education. In the first place, knowledge about
school must go to the parents and through them to the general
public. I would like to refer to the reform movements that are
taking place in Australia. They are organizing summer courses
for teachers and pupils, and these courses arouse larger and
larger interest from year to year and get publicity in the press,
radio, and television. The full program is televised twice a day,
in the morning and the evening and is being carried on under the
watchful eye of a public of three and a half million people. This
is a real way to raise the interest of the pupils and through them
the general public.

   A number of people addressed their remarks to the question
of why students were avoiding science and turning to other areas
of specialization:

   Mrs. M. B. Palma-Vittorelli (Italy): I think that perhaps the
main reason why the young generation is now running away from
scientific education is that they are of the opinion that science
has contributed to increasing rather than solving the problems of
society. Up to now scientific education has been mainly concerned
with the problems of preparing more physicists, more engineers,
more people for the technological needs of society. If the new
generation now does not believe in the technological society, it
rejects science. So what I think is the main aim of science education at the moment is to give the young generation the kind of
preparation that tells them what the fundamental point of science
is: how to state problems clearly, how to make observations and
to make use of them, how to work from observations to their consequence, how to interpret facts, and so on. We can really reach
this goal in teaching science, and it will be useful for the minis-

ter of state, and the revolutionary as well as for the social
worker.

E. Nagy (Hungary): Why is it that so many young people are
turning away from physics and finding interest in humanistic
studies? I think it has a lot to do with time. If somebody wants
to be an experimental physicist, he must master a lot of laboratory techniques to be properly knowledgeable in his subjects. By
the time he is through he will have reached an advanced age. It
is much easier to be a theoretical physicist. An experimental
physicist may not be able to master all his techniques until the
age of 40 to 45. Pure theoretical physicists will have important
results much earlier, say at age 25. But for humanistic studies,
the young people feel results can be had at a much lower age, say
18 to 20. I think that the full productive capacity of a scientist
must be shifted toward a much earlier age. Unfortunately we try
to cram much of the old material into high schools in order to
save some time in the university education. I think this cannot be
done. The whole problem must be looked at in quite a new way,
and fundamentally we must arrive at a way so that a young scientist can reach his full capacity at a reasonably early age.

Several members of the Conference gave their ideas concerning the general problem of the interaction of science and society.

D. Sette (Italy): The subject is not so much what we can do
to increase the number of people going toward science, or how
to try to change the training which we now have. There are needs
of another kind, needs to prepare a new generation. Recall what
Professor Kapitza said. He advocated an education in which creativity of each individual is made important. One should, I believe, stress what science education may contribute to the basic
formation of a person. We are speaking here about the secondary-
school level, not about the university level. Here contributions
may be made to the formation of character, intellectual honesty,
and the possibility to make a certain set of assumptions and yet
to be ready, if somebody else proves these assumptions wrong,
to change them. These are what the study of science can show.

J. A. Rodriguez (Venezuela): Many years ago it was relatively easy to be a successful teacher. The teachers taught some

truths and it took such a long time to change that the pupils did
not notice any change during their whole lives. The truth was
good enough to assure a successful way of living and to do successful professional work. Today progress is growing so fast
that what the teacher teaches is in danger of being overcome by
progress in a year or two. When this happens, the pupil begins
to doubt the teacher, and it is for this reason that there is a
shortage of faith among the pupils today. If we want to contribute to solving the problem, we must try harder to change ourselves, be less dogmatic in our education, and keep the teachers up to date.

E. M. Rogers (U.K.): The young person asks "What use is
physics to me? What value does science have for me?" and we
find young people ignorant of those values. But I think we must
blame ourselves. We need to be advertising men. Remember
that a capable advertising man can make families wish to buy
two refrigerators when they do not even need one. We do not
advertise our physics. In reply to the question, "What use is
this to me?" we say to students, "We will drill you in Newton's
laws of motion" which will not help him drive a car or understand television. We need to review the content of our physics
teaching very carefully and to look for those things which we can
tell young people will be relevant. The samples that we offer in
school lack advertising value. I think we need to put our own
thoughts in order about what we teach, as well as on the broader
scale of the relevance.

A. V. Baez (U.S.A.): I think it is necessary, before we can
proceed to answer the question "How can secondary-school
teaching be made responsive to the new needs of modern society,"
to specify what we think are the needs of modern society. Only
then may we ask what secondary-school teaching can do in this
situation. I call the serious problems of modern society the
four P's: Population, Pollution, Poverty, and Peace. What is
the relevance of education in general to these problems? I think
it is because our young people sense intuitively, rightly or
wrongly, that physics does not seem to have any relevance to
these problems that we are in trouble, or perhaps it is that the
relevance has not been pointed out to them. I think that for this
reason young people are rebelling and dropping out of the study
of science in general, and physics in particular.

I was a little disappointed that some people say, "What we have to do is to increase enrollments in physics and therefore we have to sell physics." I believe that this is not the direction young people are seeking. Actually the young people are using the experimental method and are exemplifying some of the best features of science, even without being scientists. They exemplify what I have called the human qualities of science, that is, the longing to know and to understand. They are questioning many things, including the validity of teaching physics, searching for data and the relationships to give them meaning, the demand for objective verification, respect for logic, consideration of premises and consequences—all these are human values of science.

What is the relevance of physics to these particular qualities which I have called the human values of science? I think that this deserves much more thought than we can give it here, and I suggest therefore that it be put on the agenda for some future conference. In such a conference we must try to define the problems of modern society and then to ask what is the role of physics and physics education vis-à-vis these problems.

The first of these problems lies in the tradition that the scientist should not concern himself with the details of society. But now the pendulum has swung, and there is considerable sympathy for a more active participation by the scientist in society. This needs to be elaborated, in particular by finding examples which arouse interest in the subject and alert students to the impact of science on society.

A second working group at such a future conference might devote itself to the invention of problems and questions for textbooks that lay emphasis on the needs of society.

Physics education journals could be encouraged to devote a section to the problems of the social responsibility of science. It might also be worthwhile to consider ways in which the image of physics might be improved in the world at large.

These are but some of the ways to infuse education with the idea of social responsibility.

F. Watson (U.S.A.): The prime purpose of this conference is to consider suggestions, additions, and modifications for the development of physics teachers. If the nature of the science to be taught in a classroom is to be modified significantly in future, then we had better get going. We at Harvard Project Physics are already encountering great difficulty with teachers who have no background in the history of science, in the philosophy of science, and have a confused idea about the nature of the laws of technology and science. The teachers we now have in high schools are the product of the educational system which we have been running. If in our countries we find that the students are unaware or confused about the relationship between science and technolgy, the world of thought and philosophy, it is because we have not communicated properly with the teachers. So I would like to direct the attention of the audience to the need for focusing as closely as we can on broadening the pattern of instruction for the formation of someone who will become a science teacher. It is necessary not only to know a great deal about mathematics or physics. The teacher is going to be working with students concerned with the world, and the teacher has to see the relevance. That means that the designers of course materials as well as the teachers must have a knowledge and sympathy not only for the student but for his concern for the world.

N. Joel (UNESCO): I would like to discuss the concept of transfer of knowledge or attitude from one field to another. I believe that in the past there has been no such problem as relevance because the objective for teaching physics was to form physicists. So there was no transfer problem from one set of circumstances to another. But as the contemporary objectives of teaching science must now also contribute to the general education of people and to be a tool for life in general, this question can no longer be neglected. Even in those cases where teaching of physics is well done, there is no guarantee of transfer to other fields in life. Let's take the following situation: Suppose we want to teach somebody to be very, very patient. What do we do? We teach him how to fish. When we have taught him how to fish, have we created a patient person? Of course not. If you want to create a patient person, you have to make him work through different situations. He has to be patient in situations where he is probably not going to be patient, and he must understand what it is to be patient. In the same way, if we want to make physics teaching relevant to life in general, we have to introduce elements into the teaching of physics situations that pertain to real life, not just physics. We keep saying that physics develops independent judgment, helps to see things objectively, and by studying it we learn to recognize the range of the validity of physical laws. But many people who are excellent physicists go out into life and generalize and extrapolate quite crazily. It seems to me to be evident that physics teaching would become more relevant only if it gets a bit more integrated with other subjects. If we are supposed to generate more rational people who would be more understanding of other people, who would be able to live peacefully, and who recognize the difference in people without disliking them, we cannot do so through teaching physics. So perhaps, unless the structure of education changes a lot, there is no answer. Maybe the whole system of adults being treated like children in our schools makes the whole thing irrelevant. By adults I mean people of the ages 18 or 19 or 22 who sit in schools like children. Maybe in future there will be a system by which people sometimes spend their time in one school, sometimes in another school, then they go to work, then back to school, then back to work, and so on. I think as long as we limit ourselves to present conditions, things will remain pretty irrelevant. Perhaps I am a pessimist!

F. Watson (U.S.A.): The things that have been underlying our concern for relating physics as a historically very ancient and admirable example of the sciences have been not only the technological applications but rather the intellectual overtones that have shaped our literature, strongly influenced our art, influenced our poetry, have been at times in the past in conflict with theological views, and are right now in conflict with philosophical views about the variety of ways by which the world may be described. There is a certain naiveté on the part of scientists when they seem to assume that the only way to describe the world is in terms of their particular set of premises. But there are other ways to describe the world — for example artistically — which are just as good a description of the world as any that a physicist creates. The trouble is that we don't honor it, yet we expect them to honor ours. This is the conflict of values which is occurring throughout the world. Students are now alerted to this possibility of alternate ways, yet we keep insisting that we have the only way.

If we are going to make any response at all to the growing awareness in the society, and try to indicate that we are not anti-

humanists, that we are people among people, we have to be willing to be a little more objective, a little more modest, and a little
more realistic about the sciences we are representing. So I see
a fundamental conflict in designing the pedagogy. It is relatively
simple to want to maintain the presentation of the theoretical
structure of physics, evolving but in a way sacrosanct. This is
what we have done in the past, and we are continuing to do at the
present time. But, except for a small fraction of the people who
end up as physicists, we are relatively unsuccessful by this approach; this is the basis of much of the social crisis.

An extreme alternative would be to propose that we abandon
what is called the logical, formal structure of physics—as has
been tried in some schools—and start with the problems of the
world. You start with problems of population or pollution, and
try with great difficulty to work backward from that into the
kinds of information which are at the moment seemingly relevant to the solution of the problem. Part of the difficulty there
is that the problems will change faster than our insights will
change. I think that it is probably not a highly productive model,
but I suggest it to you simply to show the range of alternatives.

E. W. Hamburger (Brazil): I think we are lingering on a
misconception. To me it is quite clear that the problems of the
four P's are not scientific problems, not technological problems—they are social and political problems. Maybe in twenty
or forty years the problems of the world will be scientific, but
that is not the situation now. We do not need any more scientific
progress at present; of course we need it in the long range so
that it is good that there will be some, but the essential problems of humanity now are not in this area. I do not know a single
country where the real problems are in the physical science;
they are all of a social, political, and economic nature. Therefore I think we must accept the fact that physics is not as important as it once was.

Another thing which I think must be quite clear to us is that,
throughout history, science and technolgy have always been for
sale. Scientists and technologists have always been servants.
They have been servants sometimes to the most horrible re-
gimes, they have rarely refused to do things which we now think
were very bad. If a certain regime has lacked scientists at some
time it has always been able to import them at any time. There-

fore it seems to me that what the students are feeling is exactly
the problem of power. When changes are necessary in society,
basically the problem always tends toward the power structure
of society.

I think that science has some relevance to society in the following way: at the beginning of history man was a slave to nature,
a slave to superstition, but during the last two thousand years he
has become ever more a master of nature and in principle is now
the master of his destiny. I do not think there is any question
that the scientific method is indispensable if man wants to determine his own destiny, but I think that is the only real relevance
that science and science teaching have to the problems of today.

S. G. Bronevshuk (U.S.S.R.): As in many countries, so also
in the Soviet Union education is determined by the social needs
of society. We teachers have the duty to fulfill this need as well
as we can. What are really the social needs today in the period
of rapidly developing technology? A greater percentage of our
youth comes in close contact with technology and with science
physics, chemistry, and biology than ever before, and therefore it is extraordinarily important to adjust the level of science teaching to the social needs of society.

In the Soviet Union there are 50,000 secondary schools and
about 50,000 eight-year schools. Together these are 100,000
schools involved with the first period of education. I would like
to emphasize that with such a big number of institutions, any
small problem can grow enormously. I think you will all agree
with me that the solutions of all small problems are alleviated
by a central government system. For instance, for new experiments we need much new equipment and many measuring devices. Our ministry of higher education has its own industry for
producing such equipment. This industry is independent of all
other institutions, so it makes it easy to cooperate with the main
educational departments.

At the moment we have some special difficulties we would
like to overcome, and therefore we have investigated very deeply
the development of cultural enrichment in education in Japan and
the U.S.A. Also, we recently obtained information about rural
schools in France and in other countries. Although we find the
information we get from these discussions with our colleagues
to be useful, there are still unsolved problems.

We feel that it is impossible not to study physics; it is the basic science concerning nature. We must teach physics not only to prevent a loss in the essential content of general education, but also because physics is connected with the formation of a basic ideology. Physics affects philosophy, and therefore all members of society must learn the main problems of the physical world.

During this conference we really have not talked at all about the significance of physics to the students while they are in the process of learning it. But we must do this, if not now then in the future, in order to give the pupils a precise view of the world. We should proclaim the objectivity of the natural and physical laws, to convince the pupils of the greatness of man, so that by learning the laws of physics he is able to take possession of nature and does not have to wait for the generosity or goodness of nature to make progress. This really has to do with the philosophy of physics, which is a matter of great importance and perhaps should be the subject of the next conference.

Chapter 2

Constraints on Teacher Education

A great deal of the time of the conferees was taken up in group discussions stimulated both by the invited papers and by the contributed papers which were distributed at the start of the Conference. During the final session of the Conference each group submitted a set of guidelines that were discussed and accepted by the Conference as a whole. In presenting these proceedings, at the beginning of each chapter, based on the results of these working groups, we will give these guidelines as submitted, followed by pertinent papers and discussions which clarify and enlarge on the working group reports.

The first of these reports carried the title Teacher Education in Physics: Limitations.[1]
Limitations encountered in the education of teachers of physics can be described under three general headings:
1.
Limitations imposed by the aims of secondary education and those which arise because of the place of physics teaching in science education.
2.
Problems arising from the expansion of secondary education.
3.
Limitations arising from the ages, abilities, and aptitudes of the pupils and from the nature of the duties of the teacher.

1.
Limitations imposed by the aims of secondary education and those which arise because of the place of physics teaching in science education.
Essentially the problems are concerned with the response of

----

1. Members of the working group which discussed this subject were: R. L. Krans (Netherlands), Chairman, Mrs. M. Ferretti (Italy), Rapporteur, S. G. Bronevshuk (U.S.S.R.), Mrs. M. Chytilová (Czechoslovakia), Mrs. M. Cordier (France), E. Eisner (U.K.), J. Fuka (Czechoslovakia), D. W. Harlow (U.K.), W. C. Kelly (U.S.A.), A. D. Pickar (U.S.A.), H. Silver (U.K.), S. Sotier (Federal Republic of Germany), V. Weltchev (Bulgaria).

physics education to (1) changes in science itself and to changes
in its presentation in schools and (2) to wider changes in the educational system and in society itself.

Physics is a necessary part of a balanced education. It is an
essential component of the range of studies generally considered
desirable as a part of the general education of all pupils; moreover, it has a unique role to play in the education of future specialists in science and technology. The following, therefore, are
important:

The place of physics and the aims of physics teaching in secondary schools must be kept under constant review.

The appropriate authorities should ensure that (1) physics be
a required part of secondary education to an extent which reflects
the position of physics in modern thought and in society, and (2)
where educational changes affecting science teaching are contemplated, such as the introduction of integrated science courses,
the training of teachers should be suitably modified.

Since the aims of education cannot be attained through the
study of a single subject, it is essential that the prospective
physics teacher respect the value of the other subjects included
in the total curriculum.

2.

Problems arising from the expansion of secondary education.

a.

The shortage of teachers has compelled some countries to reduce the quality of physics teaching in secondary schools. This
situation should be rectified as soon as possible, either through
additional training for present teachers, by improvement of the
training of future physics teachers, by more vigorous recruitment, or by some combination of these.

b.

A particularly important approach to the problem of the teacher
shortage could well be the encouragement of girls to study physics, especially where physics is an optional subject. In any event,
physics should form part of the secondary school education of all
students, regardless of sex.

c.

In some schools, the recruitment of a teacher who is a physics
specialist cannot always be justified. It is recommended that science teachers receive training in several sciences in addition

to their speciality so that they may be able to conduct classes in several subjects.

3.
Limitations arising from the ages, abilities, and aptitudes of the pupils and from the nature of the duties of the teacher.

It is evident that the teaching of physics, and therefore the education of physics teachers, must take note of the ages, the abilities, and the aptitudes of the pupils.

The nature of the teacher's duties has a direct bearing on both his ability to teach effectively and on the image of the physics teaching profession. The most important consideration in this respect is the need to avoid overburdening the teacher. Eighteen hours of class work including laboratory time per week should be considered a desirable maximum.

Five problems require special attention:

1.
When specialist physics teachers teach other laboratory-based subjects, it is essential for them to have adequate preparation time if they are to cope with the disparate elements in their teaching.

2.
Wherever there is a trend toward the teaching of integrated science courses, physics teachers should be given sufficient time and opportunity to develop their knowledge and interests in related fields.

3.
In those countries where physics teachers do not have adequate ancillary support, such help should be made available as a matter of urgency. One technician per school physics department should be considered the basic minimum.

4.
Constant advice and information should be available to teachers with regard to the selection of laboratory equipment.

5.
The physics teacher must be given time and opportunity to do creative study and work in his subject, including work of a pedagogical nature. However, this should not be expected to intrude excessively on his personal life. He must also be given realistic opportunities to keep up with the latest developments in physics.

This is crucial for teachers who live in small provincial towns.

Background material which led to the specific guideline that

IT IS IMPORTANT THAT THE PLACE OF PHYSICS AND AIMS OF PHYSICS TEACHING IN SECONDARY SCHOOLS BE KEPT UNDER CONSTANT REVIEW,

was contained in a paper "On-going Curriculum Development and the Problem of Achieving it in the U.K." by M. Underwood (U.K.):

The tradition of practical science teaching in schools in the U.K. is a long one. Over one hundred years ago a committee of the British Association for the Advancement of Science was searching for "the best means of promoting Scientific Education in Schools." They drew the "important distinction between scientific information and scientific training."[2] Fifty years later laboratories had been established in many schools and science was an essential part of the curriculum. Yet problems remained. In 1918 a committee headed by J. J. Thomson reported that "a great part of the difficulty arises from the fact that the teachers, from lack of training and of knowledge of the work of other teachers, tend to go on teaching as they were taught themselves, and thus the work become stereotyped." Today only a small minority of teachers remain untrained, and yet an "Enquiry into the Flow of Candidates in Science and Technology into Higher Education" 1968 recommended: "There is an urgent need more rapidly to infuse breadth, humanity, and up-to-dateness into the science curriculum and its teaching."[3]

It seems as though no matter what steps are taken to improve the teaching of science, more always remains to be done. Why is this? The reason, and I take this as a basic premise for this paper, is that both science and society are changing rapidly. Our educational system and the schools within it are ill-equipped to adapt to this changing scene. The purpose of this paper is to

----

2. Natural Science in Education. His Majesty's Stationery Office, London, 1918.
3. Council for Scientific Policy. Enquiry into the Flow of Candidates in Science and Technology into Higher Education. Her Majesty's Stationery Office, London, 1968.

examine present procedures for facing this problem and to suggest a restructuring which might lead to an improvement in the
situation.

Ideally, one would like to see every individual school adapting its curriculum to the needs of society and its pupils, use-
fully contributing to the life of the community and adjusting to
the implications of advancing frontiers of knowledge. Unfortunately, the typical secondary school (with pupils of from 11 to
18 years of age) is too small to ensure spontaneous organic
growth from within.

However, desirable as natural advance within individual
schools may be, if it is unlikely then we must look to external
agencies to achieve it. Perhaps the preparation of teachers
provides a way of doing this. There are two basic approaches
to the preparation of teachers: initial courses in a university or
teacher's college, or apprenticeship schemes (that is, in-service
training provided by established teachers). Either of these can
serve the necessary function of initiation, but neither used in
isolation can act readily as an agent of innovation. The staff in
institutions concerned solely with initial courses, even though
recruited from schools, tend after a few years to become remote from the school situation. Anyway, no course could hope
to prepare a teacher for the inevitable and unforeseeable changes
which will come during his career as a teacher. Apprenticeship
schemes could never be accused of remoteness, but they, by
their very nature, are conservative in operation. They would be
highly satisfactory in maintaining a static condition, but such a
static condition does not exist. Various other procedures have
been tried during the last few years to bring about the necessary
changes. Of these, the many curriculum development projects
have had the greatest impact. It is interesting to note that the
first of the recent Nuffield Foundation projects[4] was in the teaching of physics. One of the virtues of these schemes has, however, led to a serious defect. They have been so thorough and
searching that the cost has been enormous. Well over one million pounds has been spent on Science Teaching Projects alone
during the last ten years. Expenditure on this scale makes it unlikely that a comprehensive study of physics will be made again

----

4. See Appendix C.

in the U.K. for some time to come. Also, such studies are inevitably slow. Events overtake the work involved. From the start of the scheme, through trials to final publication takes at least five years. The first Nuffield physics project was restricted to providing teaching materials books and resources for pupils from 11 to 16 years of age in the top 25 percent of the ability range.[5] This seemed sensible in the early 1960s when secondary education was arranged on a selective basis. By the time the materials were published, however, the country was in the throes of a change to comprehensive (nonselective) education.

Another guideline which was accepted by the conference stated that

IT IS IMPORTANT FOR THE APPROPRIATE AUTHORITIES TO ENSURE THAT (i) PHYSICS BE A REQUIRED PART OF SECONDARY EDUCATION TO AN EXTENT THAT REFLECTS THE POSITION OF PHYSICS IN MODERN THOUGHT AND SOCIETY, AND (ii) WHERE EDUCATIONAL CHANGES AFFECTING SCIENCE TEACHING ARE CONTEMPLATED (FOR EXAMPLE, THE INTRODUCTION OF INTEGRATED SCIENCE COURSES) THE TRAINING OF TEACHERS BE SUITABLY MODIFIED.

In a paper "On 'Vertical Links' Between School Physics and Advanced Physics," O. Eisler (Denmark) suggests that a school course reflecting a modern view of physics itself should make use of the vertical links between the simple problems of the school laboratory and the advanced discussions of the professional physicist. He asks:

Is it sufficient to train the physics teacher essentially at high school level more thoroughly and more extensively than in high school itself but not in a basically different way nor with a different outlook, with the explicit goal of providing a "good working knowledge" of the subject to the extent required by the presently accepted school syllabi? Or, on the contrary, is it necessary to embark upon a kind of "maximum program" of training, which

----

5. Nuffield Physics. Teachers' Guide. Longmans, Penguin, London, 1966.

would impart to the teachers a knowledge of the basic ideas and fundamental principles of modern physics at an advanced level, including quantitative-deductive theories, and even a minimum of calculational skill far beyond the immediate requirements of school curricula? As far as I can see, there are two essential arguments in favor of the maximum program.

For the first, the explosive development of science and technology has blurred the boundaries between "school physics" and "university physics," or even "frontier physics." What was (and still is) frontier physics in the forties and fifties, such as the theory of semiconductors and semiconductor devices or the unexpectedly rich spectrum of elementary particles, has become standard university physics in the sixties, and is already penetrating into the high-school curricula. The same is true of such advanced theoretical concepts as probability waves, quantum states, the connection between symmetry properties and conservation laws, not to mention such "straightforward" topics as the special principle of relativity. This implies that the teacher must be prepared to teach not the physics of yesterday but rather that of today or perhaps that of tomorrow. Good theoretical foundations in modern physics are therefore essential.

For the second, creative teaching, which requires a flexibility of approach and exposure, an ability to treat a particular topic in alternative ways on a wide spectrum and to shed light on a problem from different angles of view as well as a capability to design a teaching program presupposes a broad horizon and a depth of understanding of the subject far beyond what is explicitly required by the syllabus and textbook.

One aspect generally neglected in university teaching as well as in teacher training is that commonplace problems, occurring naturally in high-school level teaching—or to put it more dramatically, physical effects and circumstances which seem self-evident because we encounter them in every minute of life—are not necessarily elementary. More often than not, they can only be explained at all at the present level of understanding physics.

Such vertical connections between elementary and advanced physics are often neglected. Either you teach elementary physics in the high school or teachers training institute, with practically no reference to advanced physics, or else you give an advanced level course, in which any mention of elementary problems is regarded as sacrilege.

However, in the teaching of physics, at the elementary as
well as at the advanced level, these vertical links must be
brought out. This is particularly important for high-school
teachers. Both in educating new teachers, and in the in-service
training of the old ones, physics courses of a new type are required. Such courses, would focus attention on the vertical links.

As a case study, demonstrating the fruitfulness of the suggested approach, we may examine the possible vertical links
leading upward, for instance from Newton's laws of motion:

1.
After a discussions of the tautological character of the law of
inertia, it can be shown that it is a consequence of the principle
of relativity.

2.
A logical analysis of the consistency and meaning of Newton's
laws, possibly along the lines suggested by Ernst Mach and
Poincaré.

3.
The "violation" of the third law in the case of a time-dependent
distribution of the forces (for instance in the case of magnetic
forces produced by and acting on moving charges). In this connection it is necessary to introduce the field concept, the "momentum formulation" of the second law, and to prescribe energy
and momentum to the field (which makes the field a physical reality, not merely a convenient working hypothesis).

4.
Demonstration of the invariance of Newton's laws with respect
to the Galilei transformations. A discussion of the invariance or
noninvariance of Newton's laws with respect to the Lorentz trans-
formation, the advantage of the momentum formulation, and the
necessity of allowing the mass to be velocity (or energy) dependent.

5.
The laws of motion in accelerated frames of reference. The
equivalence of inertial and gravitational forces. The general
principle of relativity.

6.
The impossibility of simultaneously measuring position and momentum in quantum theory. Discussion of the uncertainty relations, complementarity of measurements, and wave packets.

7.
The limited validity of Newton's laws of motion. The classical
limit of quantum mechanics. The breakdown of classical approx-
imations and classical pictures in the case of force fields rapidly
varying in space or time.

Courses of teacher education in physics of such a type would
be concerned with an understanding of physics and of the limitations that are inherent in the choice of a particular set of laws
and axioms.

Describing a college course in physics, chemistry, and biology as a preparation for secondary-school teachers, A. D.
Pickar (U.S.A.) points out:

A program of a multidisciplinary nature cannot, of itself,
provide a totally adequate background for a teacher of physics,
chemistry, or biology. On the other hand, on occasion these
subjects are taught by people with less formal exposure than is
provided by such a course. This is more likely to be the case in
rural areas or in developing countries. Teachers are trained in
one science or another with little or no exposure to other sciences
in which they may one day be called upon to conduct classes. An
intensive multidisciplinary course is not proposed as a complete
background for secondary-school science teaching, but as a uniform requirement in training it might provide sufficient exposure
to enable a teacher to develop competence when it is needed.
There are of course many teachers who receive training in
several of the sciences through conventional courses. There are
a number of advantages of a high-quality multidisciplinary course
over this approach: there is more opportunity for a prospective
teacher to make a rational choice of his ultimate specialty since
he is exposed to several sciences early in his training; the various sciences are presented to him in relatively small blocks of
material, so that he never gets hopelessly behind in any one subject; some time is saved because those topics in which the material from one science overlaps that of another need not be discussed more than once; he is better able to see the significance
of other sciences for his own special area, thus providing him
with valuable insights for dealing with students of varying inter-

ests; and finally, it gives him an appreciation of interdisciplinary studies which is likely to be of interest to secondary school students.

L. R. B. Elton, P. J. Hills, and S. O'Connell (U.K.) in a paper describing an investigation into "Self-Teaching Situations in a University Physics Course for Secondary School Physics Teachers" state:

An earlier study[6] into the problems facing students in the transition from school to university, indicated that students needed most help in situations in which they were not in immediate and direct contact with a teacher. This in turn led to the realization that it was the learning situation rather than the teaching situation that should be the starting point for any investigation, and finally that innovations in teaching should spring from the demands of the learner. In retrospect, all this seems very obvious.

It is commonly said that the student must learn to teach himself, but he is rarely given much help in this process. With the development of modern audiovisual aids and programmed instruction, and the acceptance of a structured approach to teaching and learning, there has come an increasing awareness that these techniques have a greater potential than was first realized and that, when learner based rather than teacher used, they can become valuable tools in the creation of a suitable environment for the learner.

A good deal of discussion centered around the statement:

ONE LIMITATION ENCOUNTERED IN THE EDUCATION OF TEACHERS CAN BE DESCRIBED UNDER THE HEADING OF CONDITIONS ARISING FROM THE AGES, APTITUDES, AND ABILITIES OF THE PUPILS.

Mrs. L. Leboutet (France) in a paper entitled "Reactions of Adolescents to Physics Teaching" wrote:
----
6. S. O'Connell. "From School to University," Universities Quarterly 24, 177, 1970.

In these days, when scientific and pedagogical problems come into prominence, it is important to know the mental capacity of our pupils and the way they respond to our teaching in order to make our teaching more efficient.

Physics is taught in various countries to children of different ages and stages of development. Their knowledge of science is determined not only by the progress of science and by the program and methods of teaching but also by the abilities of the pupils, who are growing toward the mental structure of the adults, reaching this stage at about the age of 14 to 15 years. The mental powers of children develop through exact phases according to strict laws, but these phases occur individually at different ages. The teacher has to know these phases, because his pupils are able to respond to the impulses of the physical world and to the teaching of science only according to their mental capacities. The acquisition of the abstract concepts of physics goes parallel to this development, and the age of 14 to 15 is an important and critical one. Before this age physical facts are formulated in terms of common sense, and they receive their conceptual character only afterwards.

The dynamic functions of the mind are inspired mostly by the motivation of interests, and scientific interests are those which are known the most profoundly by the psychologists. The exploration of these interests at the earliest age is indispensable for the education of the physicists of the future.

The semantic difficulties for pupils of 13 to 15 years in the early stages of physics education and the consequences for teacher education were described by R. L. Krans (Netherlands):

In the education of physics teachers, attention must be given to the special difficulties experienced by average, young pupils in learning physics. For physics has its own barriers to entry for many students. In particular, semantic barriers are not given sufficient attention. A consideration of these difficulties should form part of the education of every physics teacher.

"What is physics? What do physicists do?"[7]

----

7. J. S. Miller. **Teaching Physics Today.** Organization for Economic Development and Cooperation, 1965.

And what <u>do</u> physicists do? They contribute to the building
up of a schematic verbal outline of nature by a confrontation of
the phenomena with their descriptions. This confrontation leads
to the continuous refinement of the formulation and to the closer
observation of the phenomena, that is, to theoretical and experimental developments.

The teacher should repeatedly draw his pupils' attention to
the semantic aspects of physics, for otherwise they will be overlooked by the average pupil.

Many teachers exhaust themselves in demonstrating examples
of "doing physics," letting the pupils take measurements, and so
on. Their attempts to make the physics lessons interesting may
have the result that the pupils are gripped by the facts, but this
will last for only a short time. The pupil may learn what the
domains of physics are, but these domains and the ways of ac-
quiring knowledge about them will not retain their interest, any
more than would the illusions of a conjurer. After their final
physics lesson they will quickly forget all they learned, for physics itself has remained unfamiliar to them.

And yet, if the teacher had been aware of certain elements
in the structure and development of physics and had placed those
aspects <u>explicitly</u> before his pupils, he might have got a more
enduring result. The pupils cannot develop the desired view for
themselves by simply learning the facts of physics, for they will
overlook the problems of verbalization because the subjects and
the phenomena of the physics lesson are as familiar to them as
the language that is used. They will learn a definition by heart
because the teacher demands it, but they do not appreciate the
necessity for it. But, without an awareness of the differing mean-
ings of such words as pressure, energy, weight, work in daily
life and in physics, they cannot learn to appreciate physics.

The teacher has to indicate the limitations which physics im-
poses on such common words and the necessity for this limitation. Although only a few words are involved, the teacher must
point them out explicitly whenever such a limitation in the meaning of a word is introduced until the pupils are accustomed to the
fact that physics has a language of its own.

The teacher may use the pictorial imagery of the children in
topics where the direct introduction of the concepts using schematic reasoning is too big a jump. Consider for example the in-

troduction of the units of dynamics. For theoretical reasons we
have to lead our pupils to the use of the newton as an important
factor in the whole scheme of units. Nowadays there is a strong
pressure to do away entirely with the older, familiar units of
weight (in the meaning of force). Indeed, in some countries such
units have been legally abolished by redefining the kilogram
force in terms of the newton. In my opinion, however, we should
make use of a local weight unit in our physics lessons, since the
pupils have a feeling for it. Such a use should be temporary in
nature, but it will provide a familiarity with the newton (which
they will take to be a little less than the weight of a 0.1 kg mass)
far beyond that provided by the rote memorization of the definition.

Another difficulty arises from the different modes of thought
and their occurrence in teaching. Without engaging too much in
the domain of the psychology of thinking, we may distinguish between two modes of thought: thinking by means of a series of
mental images or pictures and thinking in an abstract, formal
way on the verbal level. The teacher of physics should be aware
of these two processes and of the relation between them. Otherwise it may be that he will be using formal reasoning while the
pupil will be seeking imaginative pictorial images. For example,
in geometrical optics we have the verbal definition of a point object — "a point object is the point of intersection of the incident
rays." In the mental picture possessed by the pupil, the point object is the point source of the rays of light. So it is possible for
the pupil to have memorized one definition but to use a different
mental picture. Then difficulties will arise, especially in cases
involving more than one reflecting or refracting surface, virtual
objects or nonexistent real-point objects. The well-known definition is no more than a veneer; beneath it the pupil's mental
processes have gone their own way, led by his primitive concrete pictorial thinking.

The more abstract mode of thought is not difficult for the pupils; what is difficult is for them to begin to use it in the context
of physics which they prefer to approach as a development of a
series of phenomena rather than by logical thought.

The teacher's own career has accustomed him to apply the
logical mode of thought to physical situations, and he may forget
that his pupils have not yet acquired this background. He must

therefore take every opportunity to persuade his pupils to progress from the mode of thought which involves mental imagery
toward that which involves verbalization. Many teachers assume
that the pupils will take this step for themselves and fail to show
them the necessity for doing so.

The mental imagery which may impede the young pupils in
their search for a correct attitude toward physics must be carefully distinguished from those imaginative representations which
a physicist uses in developing his ideas. In the latter case it is
a primary logical train of thought which evokes the imagery.

A consequence is that the teacher and his pupils have a different concept of "understanding." A child may think he has understood a problem if he can relate the phenomenon involved to
something with which he is already familiar. He may then place
the phenomenon into his pictorial world with appropriate surroundings. But the teacher considers the problem understood if
he can connect that phenomenon logically and verbally with other
physical phenomena.

Physics teaching must bring the pupils from the pictorial
representations to the logical descriptions; it should arouse the
need for logical, schematic thinking in matters of physics, a
need usually lacking in young pupils. For example, no child asks
why an object falls. It would surprise him if it did not. The teacher may ask him for a cause; he expects the child to be interested
in and surprised at the phenomenon of free fall. The child learns
by heart that the object falls because it is attracted by the earth.
He may not be convinced; he has only learned an assertion. But
for the teacher whose mind is imbued with the existence of the
law of inertia, the search for the cause of free fall is a self-
evident need; he is thinking schematically.

Why is it so difficult for our pupils to accustom themselves
to thinking in this way? Because the phenomena of physics are
readily imagined to be part of a "pictorial" world. In such fields
as mathematics and grammar, where the subject matter is already abstract in nature, the pupils accept the necessity to think
schematically. They are willing to think in this mode, but they
are not usually required to do this in the physics course. This
easy escape into pictorial imagery inhibits the development of
logical thinking. The nature of the difficulty is more psychological than logical. The teacher has to see that the problem exists,

and then he may carefully and gradually remove it. He must not
think that this can be done by making a simple remark about it
to this pupils.

In the training of physics teachers, special attention must
be given to the interplay of pictorial activity and schematic thinking in the minds of young, average pupils. This should not only
be done in the general courses on child development but concretely
in courses aimed at teachers of physics where we may deal with
the question, "How does the mind of a child digest our teaching?"

IN THOSE COUNTRIES WHERE PHYSICS TEACHERS DO
NOT HAVE ADEQUATE ANCILLARY SUPPORT, SUCH HELP
SHOULD BE AVAILABLE AS A MATTER OF URGENCY. ONE
TECHNICIAN PER SCHOOL PHYSICS DEPARTMENT SHOULD
BE CONSIDERED A BASIC MINIMUM.

K. Hinst (OECD) in his paper on "Educational Technology"
points to the importance of ancillary staff in the wider field:

The reorganization of the learning process will call for new
personnel, for which no provision has been made to date. It is
therefore necessary when introducing a new learning system to
take into account whether or not such personnel can be recruited.
For the time being, this has to be done on an ad hoc basis, as
there are as yet no set qualifications for these new jobs. Depending on the learning system itself, one might need a technician or
a media specialist who would take over the maintenance functions
for the hardware, or possibly a teacher aide who would assist the
teacher and take over some of the routine work for which less
qualified personnel are required. In the United States, where the
introduction of individualized learning systems is at present more
advanced than in Europe, it is common practice to recruit these
aides from within the community and to provide the training on
an on-the-job basis. Individualized teaching calls for a greater
amount of work, and the need for auxiliary personnel becomes
even greater. So far, three special forms of such personnel can
be distinguished: media technicians, multimedia librarians, and
teacher assistants. Proper training systems become available
on a large scale.

The statement that

CONSTANT INFORMATION AND ADVICE SHOULD BE
MADE AVAILABLE TO TEACHERS WITH REGARD TO THE
SELECTION OF LABORATORY EQUIPMENT

seemed obvious to everyone, but an interesting example of one
solution to this need is provided in a description of The Centers
for the Supply of Scientific Materials by J. Cessac (France):

The Center for the Supply of Scientific Materials (CEM) is
directed by the National Institute of Pedagogy. It supplies every
French high school with the material necessary for teaching and
with chemical and biological equipment.

The organization for supplying material and equipment for
physics teaching is as follows:
1.
The improvement and selection of apparatus and equipment necessary for teaching.

The selection is made by a committee consisting of physics
teachers who are in charge of a high-school laboratory.

The catalog of apparatus, with the necessary technical descriptions, is included in a publication sent to the manufacturers
at the beginning of each school year.

Every year the teachers comment on the quality and any
faults of the distributed material, and make proposals about any
new purchases that may be necessary. In this way the catalog is
enlarged.
2.
Purchasing.

The selection committee examines the material offered by the
manufacturers and marks the items on a point scale according to
quality. Only material of the best quality is bought, and in the
case of equal quality, the cheaper item is selected.
3.
Distribution of the material among the high schools.

In each academy, the rector distributes the funds provided
by the ministry for the high schools, according to the needs.

The director of the laboratory uses the funds at his discre-
tion, being well informed about the range, the cost, and the advantages and disadvantages of the equipment.

Chapter 3

The Recruitment of Teachers

The report from the working group on the Recruitment of Teachers[1] was presented to the Conference in the following form:

In order that the teaching of physics in the secondary schools of the world may continue and improve, it is necessary to select with care those people who will eventually become secondary school physics teachers. The deliberations of the Working Group of the Eger Conference which was charged with the consideration of the problem of recruitment have been directed toward a broadly inclusive definition of the problem. Recruitment has been taken to mean any and all things having to do with identifying and enlisting those people who ought to be secondary physics teachers and retaining and improving those teachers who are already in service.

The Problem

With the exception of France, where teachers are plentiful, and the German Democratic Republic, where central planning continues to provide an adequate supply of teachers, the countries represented in the Working Group — Czechoslovakia, Ireland, Spain, U.S.A., and the Federal Republic of Germany—report a general over-all need for more and better qualified teachers, with the most notable shortages being reported from Ireland, U.S.A., and the Federal Republic of Germany. The probable reasons for the short supply of teachers include:
1.
Relatively low salary; beginning salaries may be only one-half to three-quarters of salaries offered by industry to persons with a similar preparation.

----

1. Members of the working group which discussed this subject were: W. Kroebel (Federal Republic of Germany), Chairman, M. R. Mayfield (U.S.A.), Rapporteur, J. C. Beaufils (France), F. Blain (France), J. Boutigny (France), J. Casanova (Spain), Mrs. J. Hnilickovà (Czechoslovakia), H. Melcher (German Democratic Republic), E. G. Nagore (Spain), S. O'Donnabháin (Ireland).

2.
Overloaded timetables; the 24 to 34 contact hours per week which
were reported do not include preparation time for laboratory work.
3.
Longer training time; in several countries one additional year of
training is required.
4.
Lack of prestige; in many countries the only really respectable
teaching positions are in the universities.
5.
Failure of the physics community to give proper attention to the
preparation of secondary-school teachers of physics; only a few
universities in any of the countries provide for the pre-service
teachers a curriculum different from the normal physics major
courses and directed entirely toward the special needs of secondary-school teachers.

Characteristics of a Good Teacher

It seems logical that good teachers should possess certain common characteristics. These should be identifiable in prospective
teachers early enough to help avoid mistakes in recruitment. The
following lists represent the thinking of the group concerning,
first, the characteristics desirable in any secondary-school teacher; and second, the characteristics which are specifically desirable in secondary-school physics teachers. All secondary-school
teachers should have:
1.
An intelligence somewhat above average.
2.
Intellectual integrity.
3.
An enthusiasm for learning.
4.
Good manual dexterity.
5.
An ability to communicate ideas easily and clearly.
6.
A willingness to work hard.

7.
Sufficient ambition to become and to remain professionally alive.
8.
An easy manner with people.

The specific characteristics of physics teachers should include:
1.
An inherent interest in technology having its origin in physics
and related sciences.
2.
More than average ability in mathematical manipulation.
3.
A general orderliness of thought and action.

The foregoing are not proposed as exhaustive lists but rather
as beginning points for further study by those who engage in the
recruitment of teachers.

Recommendations

The following recommendations are based upon the apparent reasons for teacher shortages, the characteristics considered desirable in good physics, and the urgent need to improve the lot
of the secondary-school physics teacher.
1.
It seems likely that students in many countries may soon begin
to take physics at an earlier age than is presently the case, thus
adding further to the need for physics teachers. Therefore it is
recommended that, as an emergency measure, teachers with a
shorter physics content preparation not be impeded from moving
to higher level teaching after further training.
2.
In many countries, the financial reward for the very good teacher seems to differ little from that for the poor teacher. In order
to provide significant incentives for persons becoming and remaining good teachers, it is recommended that the education di-
visions of the physical societies of the world form national committees of their members to devise ways of identifying good teachers and to make suggestions for rewarding them suitably.

3.

It is further recommended that:

a.

Provision be made whereby a prospective teacher can graduate
in the same length of time as any other science student.

b.

Strong efforts be made to reduce the number of hours which teachers of physics must spend with their classes. The procurement
of technical assistance is urged as one method of implementing
this recommendation.

c.

Sufficient apparatus for teaching purposes be made available at
once; this must be updated continuously to take advantage of new
curricula and new methods as these develop.

d.

A portion of the time of physical society meetings be set aside
for conferences of university professors, teacher trainers, and
secondary teachers, in which "new" physics and mutual problems may be discussed. Such conferences would serve to retain
and to improve teachers of secondary physics —an important
phase of recruitment.

   It is recognized that the recommendations listed under point 3
require the very active support of physicists if efforts to make
physics teaching a more desirable career are to be made fruitful.

4.

Finally, it is recommended that an appropriate international organization, possibly UNESCO, provide physicists with up-to-date
information from the various countries of the world. Such information should include

a.

the average number of class hours taught per week,

b.

the average number of students per teacher,

c.

a comparison of the beginning and maximum teacher salaries
compared with equivalent industrial positions,

d.

the existence of a national teacher's examination,

e.
recruitment methods.

The urgency of the very serious situation that faces several
countries, notably Ireland, U.K., U.S.A., and the German Federal Republic, with regard to the recruitment of physics teachers for service in secondary schools is illustrated by a report
from the Royal Society (U.K.) which was presented to the Conference.[2] Among the conclusions reached by the working party
responsible for this report were:

1.
Expected increases in the secondary-school population and the
increasing need to give all pupils more teaching in science and
mathematics indicate that urgent measures need to be taken to
ensure that the stock of teachers qualified in these subjects increases substantially in the next ten years.
2.
The teaching of science and mathematics is now subject to a
"vicious-circle" effect, whereby fewer good teachers are inspiring fewer pupils to study science and mathematics to become qualified to teach.
3.
Acute difficulties in obtaining funds for modern equipment and
technical assistance have an adverse effect on the morale of the
science teacher, weakening both his intention to stay in the profession and the quality of his teaching.
4.
The working party became convinced that there would be no major improvement in the situation without a real improvement in
the career structure for secondary-school teachers. In the com-
petition for the limited number of science and mathematics graduates, the career prospects offered by schoolteaching are un-
questionably less attractive than those offered by many other pro-
fessions. The career secondary-school teacher of science or
mathematics, on whom so much depends, normally reaches his
maximum salary at a point in his mid-thirties when persons with

----

2. **The Shortage of Mathematics and Science Teachers in Schools**.
The Royal Society, London, 1969.

similar qualifications in industry, commerce, government service, university teaching, and further education may confidently
expect to go on receiving increments for at least another ten
years.

To emphasize what might be done in the face of the serious situations in many countries, one specific example was offered by
M. R. Mayfield (U.S.A.) in a paper called "Physics; The Program for Teachers. Pre-Service Preparation of High School
Teachers."

   As an example of this emergency there are fewer than ten
teachers in Tennessee who meet the state requirements for certification in physics, with the remainder of the physics teachers
falling into the following categories: (1) chemistry majors with
one year of college physics; (2) biology majors with some chemistry and less than one year of physics; (3) mathematics majors
with one year or less of physics; and (4) nonscience majors with
no physics training. Further, there is little evidence to show
that the various retraining programs for in-service teachers
have provided better instruction for high-school physics students,
and the steady decline in the percentage of high school students
who take physics has not changed.
   "Physics: The Program for Teachers," now in its second
year at Austin Peay State University in Clarksville, Tennessee,
provides pre-service teacher preparation that seeks to ensure
that graduates of the program will be (1) securely grounded in
the fundamentals of physics; (2) sufficiently aware of curriculum
changes to benefit from them; (3) prepared to discuss current
ideas in physics with their best students; (4) aware of the various sources of apparatus, how to buy it, how to use it, and how
to care for it; and (5) prepared to teach in at least one other field.
This experimental program has as it objectives to develop and
implement (1) a curriculum designed for undergraduates who plan
to teach physics as a major field; (2) a method of assuring a continuing supply of good students to take these courses; (3) an unusual summer program geared to provide early teaching experience; (4) a complex of ten high school centers, well equipped and
and well staffed, where student teachers can receive excellent
experience; (5) local resources for follow-up assistance to in-

service teachers; (6) significantly increased probability that a graduate of the program will actually teach; and (7) a method for continuous evaluation of the program itself and of the performance of the teachers participating.

No easy way has been found to bring large numbers of good students into the program. We have 25 prospective teachers in a total university enrollment of 3500. Indeed it has not become completely clear how one can identify a good teacher student early enough to make recruitment of high-school seniors effective. This is not particularly surprising in view of the poor image shared by high-school physics and physics teachers in the service area of the university. Our best results have come from the efforts of enthusiastic students already in the program, who have persuaded other students presently enrolled at Austin Peay University to come by and discuss the program with us. Even then we had to learn that prospective teachers react negatively to recruitment methods that have long provided us with good physics majors. We have changed our tactics and have noted some progress, but we strongly believe that any marked increase in numbers must wait until the graduates of our program and similar programs bring new vitality to high-school physics teaching. It is toward this end also that we are assisting high schools in the vicinity of the university in improving their physics programs by providing them with consultive services, visiting lecturers, apparatus and book loans, and opportunities for students and teachers to visit freely.

Graduates of the "Program for Teachers," and other new programs will enter their classrooms well prepared and professionally alive. However, experience has shown that well prepared, professionally alive new teachers have in the past been unable to maintain themselves for any extended period of time. One major reason for this near-exponential decline in teachers remaining up to date appears to be the failure of universities to provide effective follow-up assistance. Such assistance, if it is to be meaningful, must meet the needs of the teachers, not just as viewed by university professors but, more importantly, as viewed by the teachers and their administrators. University centers to provide both demanded and center-initiated services for teachers must be developed locally throughout as much of the world as is feasible. Such a system of centers, cooperating through systematic intercommunication of ideas and activities could deliver a concentration of intellectual energy sufficient in magnitude to solve and to implement the solution of the major problems of secondary-school physics education, not merely for the present but as they arise in the future. If there are enough of us who are willing and able to develop successful pilot centers, others will follow, and physics education at all levels will prosper.

Chapter 4

Initial Training of Teachers

A working group brought together to discuss pre-service formation of physics teachers produced the following:[1]

Guidelines for the Initial Training of Teachers

Secondary-school physics instruction occurs usually at three levels, each of which may extend over different time periods in various countries. In some countries the third level is postponed to the university. All three levels should be taught by a phenomenological approach based on student laboratory work.

The first level should lead the student from observation through measurement to the basic concepts of physics and to the recognition of simple relationships between physics and the real world. The course is therefore semiquantitative in nature. The second level is more quantitative and at a higher degree of abstraction. It introduces the use of deduction as a formal method while preserving the laboratory approach. The third level uses elementary calculus, leads to a knowledge of the interrelation of laws to form theories, and begins to show the wide applicability of general principles, such as conservation and symmetry.

These characteristics determine the goals, institutional frameworks, and methods of secondary-school physics teacher education.

General characteristics of the teacher-training program. Before a student begins any program he should undergo personality and

----

1. Members of the working group which discussed this subject were: R. N. Little (U.S.A.), Chairman, H. F. McMahon (U.K.), Rapporteur, F. Balkema (Netherlands), G. Brogren (Sweden), J. B. Cross (U.S.A.), O. Eisler (Denmark), P. Fleury (France), G. G. Gluck (France), H. Hänsel (German Democratic Republic), K. Jupe (German Democratic Republic), Mrs. L. Leboutet (France), A. Loria (Italy), R. J. Miller (U.S.A.), Mrs. M. B. Palma-Vittorelli (Italy), Mrs. D. Stachorska (Poland), Mrs. G. S. Tarasjuk (U.S.S.R.), P. Thomsen (Denmark), P. Youngner (U.S.A.).

aptitude tests to guide him toward his most appropriate career and interests.

The education of a physics teacher should include a broad range of experience not only in physics itself but in the relation between physical science and culture in the most general sense. In particular, courses should stress to the teacher the relevance of science to the problems of society and lead him toward a scientific view of the world.

Teachers of secondary levels (ages 16 to 20) should be qualified with a minimum of four years of preparation, including 500 hours of physics designed for teachers and past the level at which they will teach. Since this is a minimum, the teacher thus prepared should be expected to continue his training after he begins his teaching service.

The teacher preparation in physics should have a character different from that taught to potential research physicists, in the sense that the theoretical content should be related to secondary-school teaching whenever possible. The methodology used in the courses should reflect that which is to be used in secondary schools. Laboratory work should help the teacher in the planning and organization of his secondary-school laboratory.

During the training program, the future teacher should be exposed to research experience and to physicists doing research. He should take courses in the pedagogy of physics teaching which are designed to complement his work in physics.

Specifics of the program. The education of the physics teacher should include a broad range of experiences. Among these should be included the following areas which are closely related to the physics program itself:

1.

laboratory planning and organization,

2.

equipment maintenance, modification, and fabrication of simple devices,

3.

design and performance of demonstrations and student experiments,

4.

use of teaching aids, such as computer-assisted instruction, programmed instruction, and so forth,

5.
preparation of test problems, laboratory tests, and other methods of evaluating student performance,
6.
evaluation of the comparative effectiveness of tests, equipment use, lecture techniques, laboratory work, and so on.

The program should include the following topics:
7.
mathematics courses that should be specially geared to complement the physics instruction (for the third level of physics, the mathematics must include advanced calculus and modern algebra as a minimum);
8.
the history of physics: the discovery of fundamental phenomena and the evolution of concepts in the historical perspective;
9.
educational psychology and sociology, including aspects of teaching activities such as the motivation of students, teacher-student relationships, the effects of tests and evaluations, group dynamics, learning processes, and the processes of perception, drawing inferences, and reasoning involved in scientific procedures;
10.
the connection of physics with other natural sciences;
11.
involvement in interdisciplinary seminars, profesional organizations, and local scientific activities;
12.
practice teaching, with increasing emphasis on the teaching activity as the training progresses.

The organization and presentation of a program of physics-teacher education such as this will require a major effort on the part of universities and other institutions in all countries.

The general characteristics of the teacher-preparation program which were accepted as guidelines for the initial training of teachers were supplemented in a document[2] submitted by D. W. Harlow (U.K.) as follows:

----

2. Teacher Training for Science and Mathematics Graduates. The Royal Society, London, 1969.

1.
Skills Related to the Design and Evaluation of Courses —abilities
to select lesson material for the achievement of precisely defined
objectives, to design evaluative techniques for the appraisal of
course outcomes in terms of stated objectives, and to modify
lesson material in the light of this evaluation.
2.
Knowledge of those concepts and principles from psychology,
sociology, philosophy, and history which are relevant to the
classroom and to the student's own self-improvement.
3.
Attitudes of the teacher toward himself, his pupils, and his job.
4.
Pedagogical Skills in the area of human relationships, such as
abilities to motivate pupils, to discern their differences, to understand their responses, and to modify one's own teaching accordingly.
5.
Technical Skills, including abilities to manage a laboratory, to
organize a practical class, to use projectors, and so on.

   F. G. Watson (U.S.A.) is the director of one of the major
curriculum innovation projects called "Harvard Project Physics." He was invited to give a paper entitled "Pre-Service Pedagogical Formation of Physics Teachers." His lecture was as follows:

   The classroom behavior expected of a physics teacher should
be consistent with the general pedagogical philosophy and approach to learning accepted in his country. Therefore, we should
anticipate some differences between the expectations in different
countries and also expect changes with time. Furthermore, another premise underlying this paper is that an effective teacher
is always learning and that explicit plans should be made for his
continual development throughout the period of his service in the
classroom. Therefore we shall need to consider the teacher's
pedagogical orientation both before and after he assumes the
teacher's role. But before we examine the particulars related to
these premises, a more general analysis of the teacher's role is
desirable. ·

The behavior of a physics teacher depends upon at least four factors and on how he applies them in the classroom. These four factors are: (1) his orientation toward science; (2) his orientation toward technology; (3) his orientation toward children; and (4) his orientation toward the learning process. Science, technology, children, and the learning process. At least the first three are part of his heritage and involve systems of values that have evolved throughout his entire life and, like all value systems, are resistant to ready change. Beware of the easy assumption that a short course in "methods of teaching" is likely to make a significant change in these values.

A teacher's attitude toward science is his selective summation of all his prior experiences with the things and events of the world, and especially of those specifically labeled "Science" in schools. If science has been presented as a body of rules and theorems that explain phenomena and comprise a complete system which must be learned for examinations, then the teacher will be a dogmatic authority who, with the aid of the textbook, knows the answers. His classroom role will be to "teach them physics." We have all seen much teaching in this manner and we know that such a teacher will stress mathematical formulations and complex numerical problems. Precision and accuracy will dominate the laboratory work, which is illogically interpreted as verifying the generalizations accumulated through the years from the efforts of numerous faceless men. Such an approach to science as a complete system of "right answers" will be dogmatic, authoritarian, idolatrous, and complete. The teacher will serve as the high priest of the new religion and will be domineering, meticulous, threatening, and nonempathic. Unfortunately, physics and the other sciences are commonly taught in this manner.

If, however, science is viewed as a continual quest to make sense out of selected phenomena of the world, stress will be upon the interweaving of limited empirical evidence and temporary trial explanations. Emphasis will be not only on what you know but also upon how you know it, and how well you know it. The difficulties and errors of the scientific greats, as well as their successes, will be considered in developing the idea that no accepted concepts were easily achieved and that each man has his personal limitations. The classes of a teacher who views science in this mode will stress the interrelations between evidence and theory,

the limitations of each, and the process by which any generalizations are formulated and appraised by the scientific community.

Note that these, or other, images of science are generated
during the total academic study of the teacher filtered by his selective perception. While these images have great pedagogical
importance, those responsible for pedagogical training must begin with the potential teacher holding on to whatever images he
has.

Similarly, the potential teacher arrives with his own con-
ception of the nature and role of technology. Only recently have
we realized that many physics teachers in the United States make
no clear distinction between science and technology. Perhaps this
condition obtains in other countries. These teachers have been
raised to see physics as a body of propositions whose importance
lay in their practical application through technology: better refrigerators, better rockets, better electronic systems, and as
epitomized in a slogan used widely in the United States, "better
things for better living through chemistry." They seem unaware
of the intellectually creative effort required to formulate any
tentative scientific generalizations. They have been cheated of
respect for and knowledge of the history of the science they are
attempting to teach. Also, they are unaware of the value systems that operate within scientific creativity and of the judgments
that have been made. Equally unfortunate, such teachers are unaware of the quite different criteria which must operate in technology when decisions are made about which of all possible devices will be produced. These teachers, who are unable to dif-
ferentiate between science and technology, perpetuate this confusion in their students, who in turn understand neither science
nor technology. This confusion is again the result of the orientation of the teacher's previous academic instruction in science in
the universities.

And what is the potential teacher's image of children? If
children are considered as inherently evil: defiant, lazy, and
destructive, then school will be operated like a prison. Rules
will be numerous and punitive, initiative will be severely restricted, and teachers will be domineering. If, on the other hand,
children are considered to be questers attempting to fashion their
own world, teachers will recognize individual differences, encourage initiative and diversity, and be supportive when confu-

sions occur. Whatever attitude a teacher has toward children between these two extreme positions, his attitude has developed
from the full range of experiences he has had both as a child in
his own home and as an adult in the society. This long shaping
has occurred before the potential teacher meets a teacher trainer.

These blunt statements have been made to remind you that a
teacher entering the classroom has been integrating many kind's
of attitudes and interpretations throughout his life, including many
years in academic courses. Brief exposure to "learning about
teaching" cannot be expected to make major or abrupt changes.
Whether we like it or not, we are all responsible for the development of teachers.

While we all spend our lives learning, it is strange that so
little attention is given to examining how we learn. Here then is
a crucial aspect of the teacher's function to which pedagogical
studies and practice can make significant differences. Because
studies of learning have progressed, we can state a series of
general propositions that can be learned and applied by teachers:

1.
The primary purpose of schooling is to facilitate certain cultur-
ally selected learnings.

2.
During this schooling the teacher must also be learning about the
learner and the learning process.

3.
Learning occurs individually; the rates and, more importantly,
the styles differ between learners.

4.
Learning occurs internally and can be assessed only through
overt behavior.

5.
The learner must have some purpose or goal that causes him to
commit his attention to the learning task.

6.
The learner must experience or anticipate some progress toward
his goals; otherwise his attention will drift to other activities.

7.
The most powerful rewards for learning are internal feelings of
growing competence and capabilities to cope with phenomena of
interest.

8.
Among external rewards, positive rewards are more potent than
are negative rewards (threats, punishment, and so on).
9.
Given the opportunity, students learn much from each other.
10.
Learning can occur without teaching, but "teaching" does not occur unless there is learning.
11.
The teacher arranges the environment or defines the task for the
learner. In extreme cases, the teacher's messages may be carried by nonverbal communication, programmed instruction, or
sequential laboratory experiences.

If these and similar premises are credible, we need no longer
naively assert that "teachers are born, not made." If we accept
the further premise that education should enable students to deal
with increasing effectiveness with their problems and their decision-making, both now and in the future when they are no longer
in school, then teaching in schools must provide students with
practice in making decisions and being responsible for the outcomes. Such a premise leads to quite different teaching procedures and roles for the teacher, and therefore to quite different
procedures and roles for the teacher and for the pedagogical development of teachers.

To make the point clearer, I wish to outline two distinctly
different, and polarized models of teaching and therefore of teacher training.

The first model assumes that the product of education will be
a "knowing" person. He will have broad, if not encyclopedic,
knowledge and recall it quickly and appropriately. Also, he will
have developed a number of skills with language and numbers and
be able to apply these appropriately. Learning results primarily
in "knowing that ...," and only to a minor degree in "knowing
how..." or "knowing why.... ." This model assumes that the
learner resembles John Locke's "tabula rasa," or an empty bottle
which is sent to school to be filled with information.

The instructional procedures consistent with this model include lectures, texts, demonstrations, laboratory verifications,
difficult numerical problems, and a high competence in mathe-

matics. Examinations will be important, require the recall of considerable information, and include some of the classical "difficult problems" on which the student may have been drilled. Success in a physics course, grades, or promotion will be appraised mainly from examinations heavy with numerical problems and laboratory reports and will often be scored for neatness as well as correct answers. Such a course will be described as "meeting standards" —whatever that may mean. All students are treated alike, with the same experiences, on a tight time schedule. Physics will be presented as an exact science, and laboratory work will stress precision and accuracy, preferably to four significant figures, even when no significance is then attached to the figures. The syllabus or textbook will have been covered, but little about the nature of science as an intellectual creation will have been uncovered.

A teacher functioning within that model needs a considerable knowledge of the current corpus of accepted results of theories, laws, experimental evidence, and logical arguments relating evidence and explanation. He also needs manual dexterity with widely assorted laboratory and demonstration apparatus which he will use or whose use by students he will direct. He will profit from instruction and practice in making careful lesson plans, preparing lecture notes, and practicing demonstrations. His evaluation of the students will be primarily in terms of their solutions of numerical problems presented in end-of-course examinations. Questions or conflicts about what to teach or what sequence to follow will be settled primarily from a consideration of the physics involved and the recommendations of distinguished physicists, whom sociologists refer to as "significant others."

According to this model, the teacher functions as an expounder, a source of information and wisdom, and a pair of clever hands. To a large degree he acts as an automaton preparing students for an examination decreed by others and uniform for all students. His pedagogical training is straightforward. His knowledge of the right answers in physics takes precedence. Skill in developing logical arguments and manipulating equipment are important. Since this teacher is making few decisions, he need not know about the history of educational ideas nor the changing arguments about the purposes of schooling. He need not know about the theory of learning. Neither need he be concerned about the creative aspects of science or its role in the general culture of its interrelation with technology. He is likely to prize the development of novel and complex questions and the creation of new laboratory or demonstration equipment. His life is relatively simple, for his perceived task is "to teach them physics."

Such a teacher will receive little joy from the day-to-day development of his students, for their results are inevitably less than perfect. Often such teachers take special pride in the occasional former student who becomes a scientist or engineer. High test scores and college acceptance are often cited as evidence, perhaps reflected glory, of effective instruction.

However, not all is lost. Some local experiences show that a start can be made toward a new orientation of teachers. At the new Stirling University in Scotland, and perhaps elsewhere, for the final three years of a four-year program potential science teachers study science, at the same time observing and teaching children. As a result, the potential teachers approach the evidence and interpretations of science in order to understand not only the accepted results but also the alternative interpretations and why they have not been accepted. Three years of working with children and science result in a new respect for both and greater ability to deal with the cognitive problems of the students.

At Harvard we have found that many experienced teachers, who have been trained in the restricted science-is-technology mode or the you-must-know-the-right-answer manner, become excited when they begin to study the history of their science and to consider some of the social implications of scientific activities. These studies bring a renaissance of the teachers which will modify their outlook and behavior for many years. It is a pity that these broader approaches are so few, or are so far apart in a teacher's career.

A second model of the role of teachers, including physics teachers, is based on quite different premises about learning and the purpose of schools. Here the focus is upon the learner, the process of learning, and the conditions under which learning is enhanced. Thus the teacher's role in interacting with students is quite different.

The basic premises of this approach are: (1) that learning occurs individually within students as a result of their activities, (2) that students differ widely in their interests and abilities, and

(3) that learning is manifested by changing behaviors. School is viewed as a place where students can learn with the help of mature, experienced, sympathetic adults. This premise recognizes the individuality of the student and the possibility of differing learning styles. Because students differ, they require a diversity of learning materials and the freedom to select among them. Direct interaction of students with materials is central; no longer does the teacher stand between the student and the phenomena. Instead, the teacher's role is that of a wise helper or guide, who is aiding the student to formulate, test, and justify his own conclusions. The fact that learning occurs is evidenced by a variety of changing behaviors, including student initiative, self-reliance, joy, independence, and creativeness, as well as competence in manipulating the major concepts of the subject. Teacher development can proceed on the same premise.

The teacher is no longer dominant. Much of the planning is done by students, individually or in small groups. Responsibility for accomplishments rests in the students, not in the teacher. Evaluation is not based primarily on tests and numerical problems, but also includes observed student initiative, creativeness, commitment. Achievement is considered more in terms of individual growth than in uniform standards of academic performance. There is no set syllabus, no single course, and no external examination.

With such an approach, the classroom becomes primarily a learning laboratory. Various investigations are proceeding concurrently, a variety of instructional media—books, films, loops, transparencies, programmed instruction as well as apparatus—are being used. Discussions, even arguments, are encouraged. Peer pressure and task orientation, rather than teacher dominance and threats, control the behavior of the students.

Enthusiastic students have freedom to proceed at their own pace, not at the pace set by the teacher for the average student in the class. Much evidence from teachers indicates that some students learn much more than they would have under tight teacher dominance. Others proceed about as expected. But all students experience a sense of self-reliance and growing competency, which is of great importance.

In such an environment it might appear that the teacher's academic competence would be of less importance, but this is not the case. When the students are free to ask questions at any time, the demands on the teacher are greater than when he schedules the topics and limits the range of questions to those on which he is well prepared. His competence is apparent not just from his memorized knowledge but also from his ability to help students learn in their own ways.

How may teachers be developed for such teaching roles? Perhaps the first action comes in the selection of candidates. If they are "physicists" who wish to expound to students, they will find the role impossible. If, however, they are interested in the learning of students, they will find the role congenial and personally rewarding.

Freeing the students also frees the teacher to consider and apply as appropriate a philosophy of education as well as one of science. To help each student with his immediate problems, the teacher must be increasingly able to observe the student, to diagnose the student's difficulty, and then to suggest suitable action. Certainly this teacher needs a firm comprehension of the psychology of the classroom which begins with clinical observations and discussions with insightful instructors and grows during a lifetime of experience.

The initial training of a future physics teacher should be only the beginning of a long-term development. At best this brief training can only alert the future teacher to some dimensions of his task, introduce a vocabulary for discussing his observations, and sensitize him to some ways of interacting with students. These are best done in clinical situations with observations and personal practice in the schools. I assume that he has previously or is concurrently studying the history of educational ideas and ideals, and also general analyses of the behavior of children and of the theory of learning. But, as every scientist knows, such bookish verbal knowledge must take on significance through personal experience. In fact, we can distinquish between several levels of knowledge and understanding: awareness, knowledge, and ultimately belief. But these cognitive abilities must relate to the realities of children in school. Also, the teacher must develop skills and decision-making powers to permit him to act effectively. These are best achieved in a clinical setting in a school where the potential teacher is sometimes the observer of others and sometimes the teacher. Some yield is also gained from the analysis of simulations, either on videotape or film. Also, small-scale teaching, brief sessions with four or five students, called micro-teaching, provides some experience and enhances the confidence of the potential teacher.

None of these procedures guarantees that the potential teacher will understand or internalize the ideas or skills being considered. Thoughtful, kindly, constructive analysis and discussions with insightful teachers and clinically oriented professors are essential. We have found that three potential teachers and one master teacher form a workable group. One of the beginners is responsible for the classroom development of all or part of a lesson planned by the group, while the others observe and then contribute to the discussion. Since the effort is to develop the abilities of the beginners to observe, diagnose, and prescribe, the discussion is initiated by the beginners. Only later does the more experienced person add his comments or raise new questions. In an attempt to ensure some immediate successful changes in classroom behavior and to lessen the tensions of the beginners, only a few (often three) specific changes in teacher behavior are proposed for the next trial. At least one change should be so specific and manageable that the beginner will immediately be successful and deserving of praise. Notice that we are simultaneously attempting to instruct—that is, increase the student's knowledge and understanding—influence his value system as to what is desirable and appropriate, and develop his classroom skills. The insights and skills develop slowly and require many opportunities for practice and constructive analysis. Videotaping of the teacher and the students in the classroom permits quick and explicit review without dependence upon memory.

Within the time generally available in the United States for pedagogical training, only a beginning can be made on the observation and analysis of classroom behaviors of teachers and students. Many important questions can hardly be raised, let alone considered to the depth needed to underlie decision-making by the teacher. Also, most curricular decisions about what, when, in what order, and with what emphasis to teach, must be postponed. This is why a long-term program of teacher development, perhaps lasting five years under sympathetic constructive criticism, is desirable. Unfortunately we find that most teachers who have had the minimal training and then been thrust into schools are hesitant, if not unwilling, to examine some pedagogical questions. Such questions appear, quite properly, to threaten their self-esteem as experienced teachers who have been reasonably successful in their teaching. One procedure, used with some success, is to have the experienced teachers act as observers and guides (while themselves under observation and advisement) to inexperienced teachers. The interaction can be highly productive to both, but much depends upon the willingness of the experienced teacher to expose himself to questions and challenges.

In summary, the pedagogical development of physics or other teachers is a lifelong operation in which many of the basic value systems have been well established before the potential teacher enrolls in any pedagogical instruction. Study of the history of educational ideas and of the theory of learning are desirable. Clinical practice and constructive criticism are essential. But the specifics and emphasis depend upon what you and your colleagues consider sufficiently important about science to be communicated to the next generation.

In an invited paper entitled "Pre-Service Formation of Physics Teachers: Technical Education," R. N. Little (U.S.A.) painted a discouraging picture of physics programs currently available in teacher-preparation institutions in the United States.

In this paper the background information comes from, and the recommendations made apply to, the educational system in the U.S.A. In this way the discussion can be quite specific. Since the lack of well-prepared secondary teachers of physics is a worldwide phenomenon, this discussion should suggest guidelines that can be adapted to any educational system. The chief focus in this talk will be on an educational system that attempts to train every individual to his limits and does not direct itself principally to the exceptionally gifted student.

Let us turn to the task of attempting to design a program for preparing teachers for secondary-school physics. First, we need to define the objectives of physics education. Second, we need to determine those boundary conditions prevailing in the secondary schools and in the teacher-preparing institutions which affect the design of the preparation program. Within this framework we can proceed. The objectives for physics must conform to those set for

education in general. A complete set of educational objectives
has recently been elaborated in the United States. An Education
Committee of the States has undertaken a project called the National Assessment of Educational Progress. One of its principal
tasks has been the definition of educational objectives. Among
these is a group of four objectives defined for science education.
These will serve very well as physics-education objectives. They
aim to enable the student

1.
to know the fundamental facts and principles of science;
2.
to possess the abilities and skills to engage in the processes of
science;
3.
to understand the investigative nature of science;
4.
to have attitudes about and appreciations of scientists, science,
and the consequences of science that stem from adequate understandings.

Many sub-objectives have been defined, but these four will suf-
fice for this talk.[3]

Focusing attention on the secondary school, it is normal in
the United States to offer physics education at two grade levels.
The first of these is usually for children of 15 years of age and
forms a part of a general science or physical science course. It
is true that some instruction in physical science occurs earlier,
even in primary school, but such instruction is dispersed and not
in an organized pattern. More than 90 percent of all secondary
students take this first course. Various sampling studies have
shown that teachers at this level have an average preparation in
college physics of less than 110 hours. The second instruction in
physics comes for 18-year-old pupils as an elective course in
general physics. Approximately 10 percent of all secondary-
school students take such a course, usually those who plan to be
professional scientists or engineers. One consequence of the
small enrollment is that most secondary teachers of physics must

----

3. Science Objectives, NAEP, Room 201-A Huron Towers,
2222 Fuller Road, Ann Arbor, Michigan 48105.

also teach a second and perhaps a third subject, as few schools can provide enough sections of physics to fill a teacher's work assignment. Another important consequence is that about 75 percent of the population gets no further training in physics than that of the first course. Included in this group is the majority of the first-course teachers of physical science!

Let us now look at the physics programs currently available in teacher-preparation institutions and first consider universities that are productive both in research and in teacher preparation. World War II clearly demonstrated the value of physicists in technological development. Forecasts predicted a severe shortage in face of an increasing need. Most university departments accepted the challenge to produce more and better physicists by modifying their courses to produce research-oriented doctoral graduates in the most efficient way. Not only were new courses developed for the university level, but the development was extended to the preparatory work in the secondary school as well.

The program evolved yields the Bachelor of Science in Physics in four years and the Doctor of Philosophy in Physics in an additional four to five years. As prerequisites it assumes a secondary-school physics course of the newer type and mathematical skills, including some familiarity with calculus. Examples of the newer type of secondary-school physics courses are the PSSC, Harvard Project Physics, and the Nuffield Project.[4] The first two years of the university program present an introduction to the major theories of physics: classical mechanics, special relativity, classical electricity and magnetism, statistical physics, atomic physics, and occasionally some quantum mechanics. A demanding level has been set for these two years by course developments at the California Institute of Technology, the University of California at Berkeley, and the Massachusetts Institute of Technology. A student is steadily under pressure to absorb established theories and to explain synthetic problems in terms of these theories. In the laboratory he is frequently told exactly what to observe, and discouraged by the pressure of time from using the laboratory as any real test of the theory he has been presented or to test alternate theories.

In the third year the same theories are generalized to cover
----
4. See Appendix C for a brief description of these projects.

more complicated systems. A great amount of time is spent in developing the mathematical tools needed for expressing the theories in their more general forms. All applications of physics to biology, earth sciences, or technology have been eliminated in favor of gaining more sophisticated mathematical skills. In many universities the student gets no laboratory experience in the third year.

During the fourth year some special branches of physics are introduced; however, the principal objective is to master the mathematical foundations of quantum mechanics. Usually this study is done before the student experimentally faces the phenomena that require quantum mechanics for explanation. The rationale of this order of procedure is that the student will have the most powerful theories now known when he begins his research observations. Thus he can make his research contributions at the highest level of sophistication.

One possible danger in using this method throughout the four undergraduate years is that the student may lose the spirit of inquiry which is the motivation for all scientific investigation. In an effort to combat this danger, most programs include an advanced laboratory in the fourth year in which student has wide freedom in choosing experimental activities. That the danger is real, however, is proved by the reluctance of many students, particularly those who call themselves theoretical physicists, to undertake the laboratory projects.

The first two years of graduate study continue the clarification and development of existing theories, usually from an axiomatic approach. Only in his doctoral research project does the prospective physicist begin to construct new models and develop new theories for himself. Only then does the experimental physicist begin to draw on techniques from chemistry, engineering, and other disciplines. The theoretical physicist probably never makes significant contact with any other discipline except mathematics.

Let us compare the physicist-preparation program thus outlined with the four stated objectives. The program is superlatively designed at all levels to help the student achieve the first two objectives; that is, to know the fundamental facts and principles of science, and to possess the abilities and skills to engage in the processes of science. Practically no effort is spent on the third objective, to understand the investigative nature of science, until the last years of the graduate program. In those years, the program emphasizes this objective very satisfactorily.

The next objective is "to have attitudes about and appreciations of scientists, science, and the consequences of science that stem from adequate understandings." The program accomplishes the first part of this objective by making the student a scientist and associating him with other professional scientists. It frequently fails to give an understanding of the consequences of science and technology, because applications to other disciplines have been eliminated from the program and because of the narrow specialization of the university research laboratory. From this comparison we must conclude that this program is reasonably effective, especially if its graduates find employment as university research physicists.

University physics departments offer other courses which form parts of programs for preparing humanists or other scientists. The most common offering consists of two alternative one-year introductory courses in addition to the introductory first-year course in the B.S.-Ph.D. physics program. The first of the two covers general physics but at a level that does not require calculus. It is used chiefly in premedical or biology programs and emphasizes only the first two objectives. The second is directed at humanists and attempts to give the flavor of physics without the problem-solving that is characteristic of the other introductory courses. In this way the course addresses itself to the last objective, except that it cannot presume a good understanding of physics as implied in the statement of the objective. The course explicitly neglects the first two objectives and generally does not attempt the third. Most courses of this type meet in large lecture sections with no accompanying laboratory experience. This format precludes any effective achievement of the third objective. Both of the courses just described are considered as terminal courses and do not lead easily into subsequent work in physics.

Hence at present the prospective secondary-school physics teacher must take his program as a subset of the courses forming the B.S.-Ph.D. program in physics. This subset should give adequate attention to all four objectives for science education. It is obvious from the description of the B.S.-Ph.D. program that no such subset, smaller than the complete program, exists. This was a conclusion first given publicity in the report of the Second Ann Arbor Conference on "Curricula for Undergraduate Majors in Physics."[5,6] A few years later, in 1967, a weeklong study conference was conducted by the Commission on College Physics to determine guidelines for secondary-school teacher-preparation programs. The report of this conference, entitled "Preparing High School Physics Teachers," contains many valuable suggestions.[7] This report has stimulated several physics departments in the United States to initiate new programs, but few have been as effective as one could wish. One reason for failure may be that the courses thus far developed still seem to the students to be too similar to the B.S.-Ph.D. introductory courses in emphasizing the first two objectives to the detriment of the other objectives. Evidence that this may be the difficulty can be found in nearly every survey of public attitudes toward science. Opinion polls show that most people believe science is an authoritarian, orthodox system that admits of no uncertainty or ambiguity. The prevalence of this attitude indicates that science is being presented in this manner in the secondary schools. Since the secondary-school teachers teach as they themselves were taught, the blame falls directly on the lack of emphasis on the third objective in the science courses of the teacher-preparation programs in the university.

A corollary of the popular attitude toward science is that few prospective teachers regard physics teaching as permitting the normal stimulating intellectual interaction with the student. Presumably this misconception is also due to the science experience in secondary school. It means however that even though a department develops an excellent program there will likely be few takers. The net imbalance has been chronic for decades and confirms that secondary-school physics is taught by people who did not prepare themselves to be physics teachers. No rapid change in attitude is to be expected.

----

5. American Journal of Physics 31, 328, 1963.
6. Reprints are available from the Commission on College Physics, 4321 Hartwick Road, University of Maryland, College Park, Maryland 20740.
7. Ibid.

In response to the problems illustrated by the case of the
United States, outlined by Little, the Conference accepted as a
guideline that:

THE EDUCATION OF A PHYSICS TEACHER SHOULD INCLUDE
A BROAD RANGE OF EXPERIENCES NOT ONLY IN PHYSICS
ITSELF BUT IN THE RELATION OF PHYSICAL SCIENCE TO
CULTURE IN THE MOST GENERAL SENSE. IN PARTICULAR,
COURSES SHOULD STRESS TO THE TEACHER THE RELE-
VANCE OF SCIENCE TO THE PROBLEMS OF SOCIETY AND
LEAD HIM TOWARD A SCIENTIFIC VIEW OF THE WORLD.

The first point made in this guideline was emphasized in a
paper by A. D. Pickar, "A College Course in Physics, Chemistry and Biology as a Preparation for Secondary School Teachers,"
which he concluded by saying that

interaction and cross-fertilization among the traditional disciplines is a feature of modern science which cannot be overlooked.
It can be argued that this feature should not be neglected in the
area of science education, particularly in the training of secondary school science teachers.

Although the first part of R. N. Little's paper was a report
on the present state of teacher preparation in the United States,
he went on to make concrete suggestions for improving the education of secondary school physics teachers:

Let us then focus attention on a program for a 200-hour minor
in physics taken over two years. With this start it is a simple task
to design additional courses so that with the minor as an entering
wedge those who become interested could complete a logical major in physics. This process, I believe, is the opposite to the
normal procedure. It is usually appropriate for a minor program.
The first year of the minor is all-important, as it must also
serve as the one year of science generally required of all prospective teachers. Since many will go no further in science, none
of the four objectives can be neglected. This first year thus offers the greatest challenge in course development.
A scientist studying a phenomenon first makes assumptions,

then observations, and finally inferences. The process is reiterative since the observations and inferences frequently show the initial assumptions to be incorrect. By using this process on many phenomena, more general relationships may be inferred. The scientist has no guarantee that his next observations will fit his established relationships. The new observation may very well force a completely new set of relationships. This understanding of the nature of science is completely opposed to the popular opinion of science as an orthodox dogma.

  The common lack of understanding of science suggests that the objective "to understand the investigative nature of science," should provide the framework for the first course of the minor program, with the other objectives having supporting roles. Two recent course developments emphasizing this objective will be helpful in pointing the way for designing our first year course. The first of these, called PSNS [Physical Science for Non-Scientists][8] uses as its theme the structure of matter and was developed at the Rensselaer Polytechnic Institute. The second is a completely laboratory- or inquiry-oriented course called IPS.[9] It was developed by Educational Services, Inc., the continuation of the group that developed PSSC for 18-year-old pupils in physics. The IPS course was initially designed for 15-year-old children in physical science; however, a version exists for a teacher-preparation program. The topics selected for IPS also stress the nature of matter and tend to include more activities identified with chemistry than with physics.

  The emphasis on the objective just given sharply reduces the number of topics that can be covered in a year's course. The character of the previous preparation of the students further limits the number of topics. The students in this course will have taken the minimum mathematics and science. What little they took was considered irrelevant and immediately forgotten. Algebra is not a useful working tool for these students; they will labor over simple arithmetic. Presentation of data by graphical means is outside their experience. The concepts of uncertainty and precision in

----

8. Physical Science for Liberal Arts Students. John Wiley & Sons, New York, 1957.
9. Education Development Center. Introductory Physical Science. Prentice-Hall, Inc., New York, 1968.

observations are completely new. Much of the time in this first course must be spent in the learning of skills ancillary to the experimental observations. You will find that when confronted with an actual physical system to study, most of the students cannot at first distinguish between assumptions, observations, and inferences.

The number of possible topics is small enough that there are several selections that would be appropriate. Subsequent courses can be used to treat topics not covered in the first year.

The first year will have the following characteristics as an ideal: No formal lectures will be given. The student will first meet a concept or a relationship between concepts through experimental laboratory activities. In a prelaboratory discussion the class can mutually agree on assumptions regarding the system to be observed. Students will then individually, or in a group, make observations on the system. In a section of 24 students there might be 12 identical systems under study. Postlaboratory class discussion of results permits each student to participate in the process of making inferences. The professor acts as a moderator for such discussions. In this way each student engages in scientific discovery just as effectively at his level as a doctoral candidate in physics does at his level.

I have tried this kind of course and can assure you that it is effective in stimulating enthusiastic student participation. There is so much individual teacher-student interaction required that it will probably fail in sections having more than 24 students. This limitation might seem an impossible barrier to widespread use; however, there is a solution. Advanced undergraduates in the B.S.-Ph.D. program, when adequately supervised and assisted, have been successful in conducting these sections. Indeed the reaction of the student assistants is most encouraging. They feel that this teaching experience complements their own B.S.-Ph.D. program in a profitable way. A suggested selection of topics for the first year could be:

1.
Activities selected to clarify the operational definitions of assumptions, observations, and inferences.
2.
The nature of a physical law.

3.
Time intervals and their measurement.
4.
Displacement, distance, area, and volume.
5.
Force and mass.
6.
Velocity and acceleration.
7.
Work and energy.
8.
Temperature.
9.
Substances versus mixtures.
10.
Heat.
11.
Interactions between substances.
12.
Evidence for the particulate nature of substance.
13.
Electrical behavior of substance.
14.
Applications of preceding topics to the student's environment.

The first two topics can be developed in many ways. The so-called "black box" of the elementary-school science program developed by the American Association for the Advancement of Science is an excellent example for the first unit. It consists of a sealed hollow box containing a few loose objects. The problem is to learn all one can about the interior from nondestructive observations. The study of this system can be done at many levels of sophistication; its study is analogous to the scientific study of the atom or the nucleus.

The second topic is included to show that a physical law may have a limited range of applicability or may hold only within particular limits of uncertainty. A convenient example is conservation of matter. Substances can be selected which apparently change in weight when mixed, so that the students are forced to set limits of credibility on the hypothesis that matter remains constant in any mixing process. Bar graphs or histograms are very useful in displaying the degree of uncertainty and require minimal mathematical skills. Topics 3 through 7 are intended to refer only to simple objects in which internal structure is not a question. These units would also avoid angular momentum and rotational motion. Topics 9 through 11 initiate the study and classification of types of matter. In topic 12 the student does activities most easily explained on a particulate model of matter, and in topic 13 electrical current and charge are discovered as consequences of some interaction of substances. In the last unit the student would make explicit applications of his new understanding to objects or situations in his environment. One example would be calibrating the speed of his record player turntable; others are easy to find.

The list of topics proposed might make the course sound like any other introductory physics course. What makes it different is the extent of participation by each student and the atmosphere of scientific collaboration that can be established. Let us compare the course just described with the four objectives of science education. The high degree of student participation is intended to achieve the objective, "to understand the investigative nature of science," and the first part of the objective, "to have attitudes about and appreciations of scientists and science ... that stem from adequate understandings." The particular selection of topics satisfies the objective "to know the fundamental facts and principles of science" by introducing a set of basic concepts that can serve as an appropriate foundation for later science courses. The objective, "to possess the abilities and skills to engage in the processes of science" must be introduced through the design of specific student activities, by various methods of presenting the results of observations and through class discussions. The last part of the objective "to have attitudes about and appreciations of ... the consequences of science that stem from adequate understanding" is chiefly emphasized in the activities of the last topic, which demonstrate applications to the student's environment. These applications should be introduced as activities throughout the course; however, their greatest impact will come from the deeper understanding possible near the end of the course.

In many universities this minor program will consist of no more than two years. The second year should follow the same format of high student activity but have a different selection of

topics. The selection should include those topics of greatest use
to other disciplines of science and technology. A suggested list
follows:

Section A: Radiation

1.
Light (principally geometric optics)
2.
Radio and microwaves
3.
Sound
4.
Nuclear and X radiation

Section B: Wave Motion

5.
Characteristics of waves
6.
Superposition and phase

Section C: Spectroscopy of Frequency Analysis

7.
Light
8.
Sound
9.
Alternating electric currents

Section D: Quantum Effects

10.
Electronical charge
11.
Critical wavelength in photoelectric emission
12.
Molecular speeds

Section E: Applications to Student Environment

13.
Examples selected for contemporary, local interest

In addition, the second year of the minor program should include
a mathematics course carefully phased with the needs of the physics course. In universities such close collaboration between de-

partments is occasionally difficult. In such a case, the mathematical skills will have to be included, after the student himself has established the need, in the physics course itself. There will of course be a corresponding decrease in the number of physics topics that can be treated.

After two years of the minor program there are two possible ways of continuing toward a major in physics. The first is a set of transitional courses that would allow a transfer into the B.S.-Ph.D. physics program. The transition should be made possible for those students who have the talent and who have developed a desire to do research in physics. None of those who make the transition and graduate from the B.S.-Ph.D. program is likely to want to teach secondary-school physics. Also, the further preparation is not particularly appropriate for secondary physics, especially for the 15-year-old student level. The second way is to give courses in physics that emphasize the physics used in other disciplines. Examples are many: the physics of musical instruments, physics problems in architectural design, geophysics, meteorological physics, astrophysics, biophysics, and countless examples from technology or engineering. Some mixture of these applications with certain fields of physics not covered in the first two years, for example statistical physics, is probably the answer. It is undesirable to specify further detail in the selection of physics topics for the third and fourth years of such a physics major except to say that topics must be selected with the complete program in mind. A different selection in the first two years should be reflected in a complementary difference in the last two years.

The foregoing program outline gives the course offerings of the major department, the physics department. In addition, the physics department will have to assume the lead in arranging for appropriate supporting courses in mathematics and other subjects which the secondary physics teacher may have to teach.

Implicit in the teacher-preparation program is greater collaboration between science departments than we have at present. An even more important collaboration, which has long been neglected, must occur between the departments in charge of the pedagogical formation portion and the departments in charge of the technical formation portion of a teacher-preparation program. A university physics professor is not adequately prepared to teach in the physics program outlined if he has had no personal experience in teaching at the secondary level and does not continue to keep close contact with secondary-school problems. He also needs experience with the testing and evaluation methods that are constant study in the professional education departments.

Stressing quite a different point of view, W. Kroebel (Federal Republic of Germany), in a paper entitled "The Training of Physics Teachers for Secondary Schools and the Dependence of this Training on the Instruction in Universities," championed the Ph.D. requirement for secondary-school teachers with the thesis in educational research:

I will talk from the viewpoint of what is happening in my country. We have a tradition in the education of secondary-school teachers. We call these people Staatsexamenskandidaten, which means candidates who get their certificates when they finish their studies with government aid; when they are finished, they are servants of the government. There is no separation between the education or training of a professional physicist and that of a secondary-school physics teacher. There are only a few special courses for people who want to become teachers in secondary schools, and these are the following.

First, they have only two subjects to study in general, a combination of mathematics and physics, and they do not have as many lectures on these subjects as those who want to become professional physicists. Also, they have a limitation on the number of hours they have to spend in theoretical physics courses, in mathematics, and so forth. Besides this, they have additional work in philosophy and pedagogy. After they have done all their work to meet the standard to become a teacher in a secondary school, they then have to finish advanced work in physics, generally in experimental work in one of the research institutes. Although the students thus come in contact with research, they need four years to do it. The preparation of such work takes a lot of time.

This was the situation I found when I came to the University of Kiel in 1946. I came from an industrial background, so I knew what was necessary for people to be educated in physical thinking, and I thought it was not good that at the university nobody worried about what was being done in secondary-school physics teaching

or even why it was necessary to train physics teachers for secondary schools. I tried to get answers to my questions and I read
the literature, but I could not find anything that would give me an
answer. Then I thought I would have to do something about it, and
I decided to undertake some research.

First I started with a new kind of Practicum, which was a
demonstration in which the students could learn to illustrate physical phenomena and the laws of physics as they would have to do
in secondary schools. This they do in their university courses.
We try to develop new experiments but we do not use industrial
equipment because I believe that the university physicists know
what lies in the future and industry does not.

I have tried to give Ph.D. research topics to the students
which specifically apply to physics teaching. Here are some examples: What is the performance of the human mind in pupils
when they are 10 or 11 years old? This is a question for real
research work. Yet we have no such investigations concerned
with the learning of physics in my country. Another example is:
Should physics teaching begin in the secondary schools when the
pupils are 13 to 14 years old? I have watched the pupils in the
schools and know that children of 13 and 14 have little interest in
mechanical things, and yet our tradition is to start physics lessons with mechanics. So I thought it should be possible to start
mechanics and physics much earlier. And I have had success in
my country since we have tried giving this material at the age of
10 or 11. In fact we have had a great success. Of course we have
had many difficulties. One of them was to convince the government to let us try it. Another was that the teachers themselves
did not believe that it was possible to give physics so early. And
so we have had to develop new lessons for the teachers in secondary schools to help them teach it in these earlier years. This was
a very great research problem, but we have started it and through
it we have learned many things.

Another criterion for choosing research projects is to evaluate the demonstration of physical phenomena according to pedagogical ideas. By doing this you will find that you have a lot of
research ideas for teachers to use for their doctoral theses. In
my country we need 10 to 15 percent of our secondary-school
teachers to have the doctor's degree, and they may do their thesis
work on what I may call "physics-pedagogical research." I have

convinced the physicists in Kiel that this is a good starting point
for the education of all the teachers in secondary schools, and
we have set up a research institute for doing educationally oriented research work for secondary-school teachers who are working for their Ph.D. degrees.

As an illustration of the type of pedagogical research carried
out at Kiel, Kroebel described his research work on children's
comprehension as related to the length of time spent on individual letters in words of varying difficulties.
With very much the same concept in mind, R. L. Krans
(Netherlands), in a paper called "School Physics as a University
Specialism," wrote:

The goal of our conferences must be to help the man doing
his job in the classroom, to improve his conditions of work so
that the teacher may feel himself a physicist. It is difficult to describe in general terms what has to be done, because the requirements for qualification in the different countries are so different.
In the following I start from the situation in the Netherlands.
Here the qualification for teaching at schools, preparing for
university study, demands the degree of doctorandus, which is
comparable with the master's degree in other countries. So the
qualified teacher has had a 6 to 8 years' university training in
physics. Therefore he may be considered a physicist like his
fellow students. After going into teaching, however, he quickly
loses his connections with university physics; he is to be considered a "has-been" physicist. If a suitable university specialism
should exist, these teachers could easily maintain connections with
university physicists. Research on the teaching of physics, on
school physics, and on the diffusion of physical knowledge could
be such a specialism. We call it "school physics," but the name
does not matter much.
Many teachers could contribute to the development of this
school physics. They should be allowed to get their Ph.D. degree
in physics, not in education, by writing a thesis on the explanation and teaching of modern developments in physics—for example
on the teaching of the theory of relativity in secondary education.
School physics as a branch of university research might also
break down the high wall that separates physics and physicists

from the outside world. At present it looks as if majoring in physics means preparing oneself for a life in the monastery of physics. This is not in accordance with the wishes of the students
when they start the study of physics. The majority of them have
chosen this study because they had liked physics in the grammar
schools. They have a broad interest in physics and sufficient intellectual capacity for digesting it, but many of them lack the
psychological disposition, creativity, ability to concentrate, and
perseverance necessary for a research physicist in a small part
of pure and applied physics. They remain interested, however,
in new developments of physics. A worker at the frontiers of physics pulls the car of physics in new territories. But behind this car
there is work for many trained physicists to change the rut into a
paved path along which a new generation can be brought to the
vanguard of physics. A better, more specialized training for the
work behind the front must be created; this is school physics.

**In amplification of the same conference guideline K. Hinst
(OECD) offered the following suggestions in a paper entitled "Educational Technology":**

The approach to a systematic reconsideration of the teaching-
learning process and the design of learning systems so far have
revealed that one cannot deal with these questions without at the
same time considering the factors that set the framework of the
teacher-learning process. Factors such as the present skills of
teachers, the nature of the accommodation available, and the
vagueness of present curricula in relation to educational objectives influence the process itself. Unless they are readjusted to
the new ways of instruction, it is unlikely that innovations will
succeed. It is here that a systematic approach will provide the
basis for more strategic thinking among educators, be it at the
classroom level, the school or community level, or at the regional and national levels of our educational systems, by focusing
more closely on the frame factors of the teaching-learning process while looking at this process at the same time.

Any strategy for implementing the production, utilization and
evaluation of new learning systems must accept the fact that the
element "teacher" is the crucial point for developing the strategy.
The introduction of modern learning techniques and styles will af-

fect the position of the teacher in the classroom in a way that will call for a reconsideration of his role and his various functions. The common denominator under which these changes can be regarded is the change in focus from teacher-based systems. At a time when teachers are becoming more and more self-conscious about their social status, such a change in focus is likely to create resentful attitudes, especially if teachers do not adequately understand the implications of the change. The fact that even today the majority of teachers still adopt anything from a reserved to a hostile attitude toward any media innovations is an unfortunate indication of what can happen. To surmount this resistance, one has to take account of this change of the teacher's role and functions in the classroom against the background of his social role and present status within the society. Any strategy which does not take this complex relationship into consideration is jeopardizing its results.

Knowledge and Skills

What does the individualization of learning and the production of prefabricated course material really mean for the teacher? Certainly not that such packages can take over the whole learning process. Even in the future, learning will be achieved by a variety of methods which assign a central position to the teacher—a position in which at least one function is considerably changed: the conveyance of information. Teachers today are still carrying out predominantly this one function, though many people (and there is evidence to support this) believe that they are least suited for it. It is precisely in this activity that media—and we include under this term anything from printed material to complicated machines—can take on a major part of the burden, leaving the teacher free to perform a truly pedagogic role. The latter has, so far, been a useful reference point on which to hang professional ethics rather than a description of reality. In an individualized and democratized teaching process, the teacher's job will be to promote the generalization and transfer of knowledge, as opposed to the mere learning of facts, and to foster creative activity. He will continuously observe the progress of each student and intervene judiciously wherever there are obstacles. His pedagogic responsibilities can thus be aimed more directly at the de-

velopment of the student's individual personality and behavior. In
taking on the role of a counselor, he must work in close cooperation with parents. To carry out these tasks, however, he will
need additional specialized training—both practical and theoretical—in social and educational psychology, in their methods of observation and analysis and in techniques of guidance.

As to the hardware side, it will be the teacher's function to
direct the various media in the classroom, but apart from this
he will have to give advice in the preparation of course material
and help curriculum designers by providing them with feedback information. His experience in the field will be indispensable for research workers. For these tasks, a basic knowledge
of educational and social sciences is essential, as is a minimum
. technical knowledge of media and some training in the handling
of course material.

Independent learning systems will probably be developed outside the school in most cases. However, it will be the classroom,
with teachers and students, where learning systems will be tried
out empirically during the development stage. Many learning systems will also be flexible enough to allow for special adaptations
and adjustments within the school. Furthermore, it is desirable
that the teacher prepare his own classroom activities in a more
rational and effective way. To assure students' progress, he will
therefore have to know more precisely than in the past what objectives he wants the student to achieve, whether the curriculum
or syllabus is adequately designed for certain educational objectives, and how to assess whether or not the student has achieved
them. Thus, for his active cooperation with researchers and developers, as well as for his own work with the students, it is indispensable that the teacher be familiar with the basic concepts
and techniques of curriculum development.

It thus emerges that, while freeing the teacher from one function, learner-centered and media-integrated teaching increases
the importance of several others: observation and evaluation of
students' progress; personal guidance of students and parents;
management and control of classroom activities; cooperation.
Teachers' fears of loss of professional status and of redundancy
would seem to be little more than the result of wild speculation
and inadequate information. However, it becomes obvious that it
is necessary to revise existing teacher-training curricula and to
develop courses for in-service training.

An example of a particular "Program for Teachers" given at the Austin Peay University in Clarksville, Tennessee, was presented by M. R. Mayfield (U.S.A.) in his paper "Physics: The Program for Teachers. Pre-Service Preparation of High School Teachers":

The Program for Teachers provides pre-service teacher preparation which seeks to insure that graduates of the program will be securely grounded in the fundamentals of physics; sufficiently aware of curricula changes to benefit from them; prepared to discuss current ideas in physics with their best students; aware of the various sources of apparatus, how to buy it, how to use it, and how to care for it; and prepared to teach in at least one other field.

One sequence is directed at helping the teachers know what to do and how to do it when they meet their high-school classes. Assuming a sound knowledge of fundamentals, we explore several systems of organizing and presenting introductory physics, and challenge the students to work out ways of adapting these systems to classroom situations that do not precisely parallel those for which the various high-school physics curricula were intended.

The major objectives of this sequence are to acquaint the prospective teachers with the philosophies, the capabilities, and the limitations of Harvard Project Physics, the PSSC course, and other new curricula as they are developed; to make sure that they learn to use all unfamiliar apparatus and equipment associated with these courses; to review and supplement portions of the traditional curriculum which properly augment the new curricula; and to prepare the prospective teacher for adapting, modifying, or supplementing the various curricula to meet the needs of the classes he teaches.

In order to keep these courses "current," we provide for our teachers the text materials, the apparatus, films, handbooks, and other aids that make up these courses, plus the space, time, and assistance for using them in comparing various treatments of selected topics through problem-solving, giving short lecture demonstrations, and taking the prepared examinations which accompany the high-school courses.

We have been pleased with student response to this sequence. They enjoy using the ingenious apparatus, the overhead transparencies, the single-concept films, and the very readable text materials. They are chagrined to discover inadequacies in their knowledge of general physics but are delighted to be able to correct them in a friendly atmosphere. They recognize the practical importance of intimate familiarity with the material they will soon be teaching. And, perhaps most important of all, they begin the metamorphosis from student to teacher which we attempt to bring about gradually over a period of years, rather than have them become the "instant teachers" which a few weeks of practice teaching is supposed to produce.

Another sequence is a wide-ranging modern physics emphasizing the fundamentals of physics upon which much of today's technology is based. These are solid courses which propose to strengthen and broaden the background of knowledge upon which the teacher may depend both for his formal and informal discussions with students.

The objectives of this second sequence are: to provide the future teacher with a better "speaking acquaintance" with three major broad areas of modern physics, namely atomic and nuclear physics, astronomy and space physics, and solid-state and low-temperature physics; to acquaint him further with the literature of physics and with other sources of recent information in physics; and to furnish him the opportunities to observe the applications of physics in industry and research.

This sequence has not been as successful as had been hoped, partly because of the difficulty in dealing with so many topics in such a short time; however, the field trips and the invited lecturers have been unusually well received, and we feel that the prospective teachers come out of this sequence more knowledgeable, more appreciative, and more confident. Hopefully, this brief introduction to the "people" of recent physics will promote additional enthusiasm and increased professional interest, leading the teachers to continue the studies which they undertook during these prolific years of activity.

A third sequence deals with the use of instruments, their care and repair, plus experience in the fabrication of uncomplicated pieces of laboratory apparatus as a direct objective of the first half of this sequence. Individual instruction in the use of hand and simple power tools is based upon the beginning level of ability and experience of the prospective teachers. A set of tools

considered necessary to maintain a high school laboratory has
been assembled for each student so that he may make his own repairs, look after his own equipment, and discover his own capabilities and limitations in fabrication and repair, long before he
is faced with routine maintenance problems in his own laboratory.

The concurrent performing of several modern physics experiments correlates well with tool and instrument use and provides additional opportunity for increasing the teachers' knowledge and understanding of basic physics.

The handicap of insufficient time for all students to proceed
as far as they wished with the various kinds of laboratory work
has been partially eliminated by making the facilities of the shop
and the electronics laboratory available for responsible students
to continue projects of interest on their own time.

In commenting on the guideline:

TEACHERS OF SECONDARY LEVELS SHOULD BE QUALIFIED
WITH A MINIMUM OF FOUR YEARS OF PREPARATION, IN-
CLUDING 500 HOURS OF PHYSICS DESIGNED FOR TEACHERS
AND PAST THE LEVEL AT WHICH THEY WILL TEACH,

S. G. Bronevshuk (U.S.S.R.) said:

The learning of physics is compulsory in all our schools.
With this goal in mind, we assure about 600 hours for a whole
course in physics in classes from the sixth to the tenth year in
a 10-year school. There are special schools for studying physics in more depth, in which there are 200 hours more for physics. Consequently there are about 800 hours for a whole physics
course. We have found that his 800 hours is enough to ensure the
desired standard for physics teaching at all levels in the 10-year
schools.

In relation to the Conference-approved guideline that:

THE EDUCATION OF THE PHYSICS TEACHER SHOULD IN-
CLUDE THE PREPARATION OF TEST PROBLEMS, LABORA-
TORY TESTS, AND OTHER METHODS OF EVALUATING STU-
DENT PERFORMANCE,

the statement of E. M. Rogers (U.K.) in his paper on "The Use
of New Examinations and of Examination-Construction Seminars
in Curriculum Revision and in the Training of Teachers" was
particularly germane:

In starting a program of new teaching—I will suppose teaching for understanding—we must give teachers assurance that examination questions that look for understanding can be constructed
and can be marked. In fact we must give them practice in techniques of making such questions, since they will need to make
their own questions which fit their own teaching. In conducting
the briefing session for teachers starting trials of Nuffield physics, I therefore offered round-table conferences or seminars to
groups of teachers on the subject of examination-question construction. At the first meeting I described the problem: "Suitable
questions are needed. Please invent one or two questions and
bring them to the next meeting. There you will be invited to publish your questions, and your colleagues will be invited to tear
them to pieces, with destrcutive—and I hope constructive—criticism. But remember, you will be able to get your revenge in
turn!" At the second meeting, a few days later, the discussion
grew more and more active. One teacher starts by reading his
question. Then neighbors start objecting, criticizing, defending,
changing the question, adding to it. After an hour we have discussed only two or three questions, because we have ceased to
be a question factory and have become a philosophical discussion
group vigorously exposing, comparing, and discussing teaching
aims.

I have come to regard examination-making seminars as the
most powerful technique for training teachers in any change of
teaching which involves a change of aims or of teaching philosophy, in contrast with changes which merely choose a different
syllabus or put their faith in different apparatus.

In connection with the two guidelines:

THE PROGRAM SHOULD INCLUDE INVOLVEMENT IN INTER-
DISCIPLINARY SEMINARS, PROFESSIONAL ORGANIZATIONS,
AND THE LOCAL SCIENTIFIC ACTIVITIES; AND PRACTICE
TEACHING WITH INCREASING EMPHASIS ON THE TEACHING
ACTIVITY AS THE TEACHER TRAINING PROGRESSES,

D. A. Tawney (U.K.) emphasized "The Role of Physics Centers in Initial and In-Service Training" in discussing the Physics Center at the University of Keele.

Of the widespread criticism currently directed at initial training courses for graduates provided by university departments of education and leading to a certificate in education, a high proportion is concerned with the apparent irrelevance and superficiality of treatment of the more general aspects of the courses (philosophy of education, psychology, sociology, and so on) and the lack of sufficient time devoted to matters likely to be of immediate use in the classroom.[10] Attempts to improve courses are frequently thwarted by the lack of a coordinated system of in-service training referred to earlier, for without such a system initial training for many teachers is final training and the temptation to include too much is overwhelming.

Another criticism of postgraduate courses concerns their apparent remoteness from the actual school situation: their equipment is far superior to that of schools, the advice given by tutors seems impracticable in the classroom, and the standards suggested as the norm are only achievable in a minority of schools. This criticism is easier to answer and, when made by students, springs to some extent from an attempt to justify their own uncertainty when entering an exacting profession. Nevertheless, what appears to him to be a divergence between the school situation and the picture portrayed in these courses can prevent a student from setting for himself high but realistic standards and maintaining these in the first years of his professional life. With the disillusionment that can occur when a student's expectation of his professional achievements and rewards is not immediately met, he may decide to make no effort to participate in the in-service training which, although uncoordinated, is nevertheless extensive and in many areas readily accessible. His initial training appears to have failed, and therefore in-service training is likely to be similarly unrewarding.

The aims of the initial period of professional training must include the development in the student of a desire to continue to
----
10. British Committee on Chemical Education. Teacher Training of Chemistry Graduates: An Enquiry. Royal Institute of Chemistry, London, 1968.

develop in his professional role once he has left his department, and a knowledge of the means at his disposal for achieving this development. These aims can be analyzed further into a system of objectives; to withstand his first terms of teaching he will need a basic competence, self-confidence, the ability to evaluate his own performance, and so on. However, it is not the purpose of this paper to examine these objectives further nor to discuss current research into methods of achieving them in which the author is involved.[11] Rather it is intended to suggest that attendance at a teachers' center during his professional training is likely to help a student achieve these aims.

In-service training should begin during initial training, and teachers' centers are a means for achieving this. The author has suggested to his graduate students undergoing professional training in the Keele department that they should attend the Center, and several have done so, some attending quite regularly throughout the two winter terms.

Although no evaluation has been made (in fact, the low number of students involved makes this difficult), it is quite clear that those who have attended the Center have benefited from it. They have commented that discussion at the Center with the practicing teachers has a relevance to the classroom situation that discussion with other students and their tutor lacks: "The sessions provide a very useful interchange of ideas and techniques which, because the majority present are actually teachers, are founded on experience rather than theory."[12] At the same time their tutor's ability to contribute to the work of the Center adds credibility to his work with the students.

At the Center students see teachers coming to terms with the innovations and difficulties that are part of the teacher's life. The relaxed and informal atmosphere helps them to achieve satisfactory relations with older, experienced teachers more quickly than in the tension-prone school situation. At one meeting of the Keele Center, students gave a series of demonstrations to the other members of the Center; the fact that they were able to contribute

----

11. J. T. Haysom and C. R. Sutton. Education in Science 8, No. 37, 41, 1970.
12. A. N. Poyner, Keele Concourse 63, 5, Keele, England, 1970.

in this way gave them confidence. A student can become acquainted with the teacher under whom he will subsequently work during his period of professional practice in the schools; this can help to ease a situation which some students find difficult. Perhaps, most important, the student in training sees that some teachers of physics find it valuable to meet together for that refreshment and development provided by the exchange of ideas between specialists which is the mark of a profession. Thus right from the beginning of his career the student sees the importance of communication; there is less chance that he will become professionally sterile through isolation.

THE EDUCATION OF THE PHYSICS TEACHER SHOULD INCLUDE A BROAD RANGE OF EXPERIENCES. THE PROGRAM SHOULD INCLUDE THE HISTORY OF PHYSICS: THE DISCOVERY OF CONCEPTS IN THE HISTORICAL PERSPECTIVE.

A report was given to the Conference on the International Working Seminar on the Role of History of Physics in Physics Education, sponsored by the International Union of Pure and Applied Physics (IUPAP) through its International Commission on Physics Education, which was held July 13-17, 1970, at the Massachusetts Institute of Technology in Cambridge, Massachusetts. S. C. Brown (U.S.A.) introduced the report by saying:

The International Commission on Physics Education has recognized for some time the problem of the lack of the use of historical material in the teaching of physics. As a result it arranged this seminar earlier this summer. It was attended by 32 people representing 12 different countries, and the members of the seminar included not only physicists and physics teachers at universities and secondary schools but historians of science and other interested people as well.

Many of the participants wrote papers directed toward the theme of the seminar, and they also compiled lists of books, periodicals, monographs, and archival collections which represented the work in their areas of the world in the history of science. The material that was presented to the seminar is being compiled into a final report under the direction of an editorial board.

The seminar produced a set of recommendations addressed

both to physics teachers and to the International Union. Those resolutions of interest to the teaching of physics are the following:

Archival Materials

Physics education can be enriched through appropriate use of information on the development of twentieth-century physics obtained through well-documented historical studies based on archival materials. We are concerned that these materials will be lost, scattered, or destroyed unless serious efforts are made to preserve them in appropriate libraries, archives, and museums, where they can be made available for scholarly use. We are encouraged by the results of such efforts now under way in several nations, but we wish to emphasize that much more needs to be done to provide the sound archival base that is essential for historical studies of recent physics. We urge the IUPAP to alert the physicists and physics organizations of all nations, including the national committees of the Union, to the need of preserving their own personal papers (correspondence, manuscripts, photographs, historical apparatus, notebooks, and so on) and similar records of their scientific institutions, journals, and societies.

A Book on the History of Physics

The seminar proposes that the IUPAP, together with the International Union on History and Philosophy of Science (IUHPS) appoint an editorial council of physicists, historians, and other persons competent in the philosophical, social, and economic aspects of the development of physics, for the purpose of planning a book on the history of physics, primarily for use in the teaching of physics.

While there are available a number of excellent books on various aspects and periods of the history of physics, the participants in this seminar (including physicists, physics teachers, and historians of physics from several countries) have agreed that there is no single book (in English or any major European language) that is sufficiently comprehensive and up to date to be suitable as an authoritative book for physics teachers and advanced students. It is felt that such a book, by stimulating interest in this

field, would raise the quality both of education and of research
in physics. It should be written primarily to satisfy what we believe is a clear need of physics teachers and physicists, and
therefore the proposed council should ensure the approval of the
international physics community for this project.

Encouragement and Assistance to Teachers

The seminar requests that the International Commission on Physics Education inform national authorities concerned with physics
education of the activities of the seminar and that they encourage
these bodies to establish mechanisms to allow and encourage
teachers of physics at all levels to use historical materials in
their courses.

It is further requested that the Commission encourage agencies and institutions responsible for the education (including refresher education) of those who teach physics, both at the secondary-school and college level, to include training in the use of
historical materials in their teacher-preparation programs. As
a first approximation, it is suggested that at least one course in
the history of science focusing on the history of physics and an
introduction to the case-study method is essential if teachers are
to be adequately equipped to use historical materials in their own
classrooms.

Translations of Historical Materials

The seminar requests the International Commission on Physics
Education to cooperate with the Commission on Teaching of the
International Union for History and Philosophy of Science and
other scientific bodies in establishing a committee on translations in the history of science.

Such a committee should act as a clearing house for exchange
of views of science teachers and historians on the desirability of
translations of particular books and articles of historical importance, and for exchange of information on plans for such translations with translators and publishers.

(Any person interested in the activities of the proposed committee are urged to communicate with Prof. Stephen G. Brush,
Institute for Fluid Dynamics and Applied Mathematics, University
of Maryland, College Park, Maryland 20742, U.S.A.)

Commenting on the formal recommendations, Mrs. J. Zemplen (Hungary) made the following remarks:

The aim of the seminar was to discuss methods for making physics teaching more interesting through the use of historical materials and to encourage closer contact between physicists and scholars in the history of physics. I would like to emphasize that for this purpose ways must be found to make it possible for physics students to receive the materials on the history of physics that are useful in physics teaching. Future physics teachers should be interested in this topic, so the resolutions of the seminar are of special interest to the present conference. A lot of interesting material was prepared for the seminar. There were very stimulating discussions after the lectures and in the working groups. This material will be published within a short time. I would like to present here a brief discussion of some of the recommendations of the seminar.

First, let me treat the problems of collecting and preserving materials from the twentieth century which should be very useful in teaching modern physics. This is mainly the problem of libraries and of archives and museums. Some efforts have been successful in a few countries. Therefore the seminar turned to IUPAP to get help in making its members realize the importance of such archival collections through its national commissions. In Hungary, for example, the preservation of the materials of our Physical Society falls under the History section of the Central Research Institute of Physics, but regular provisions for the collection and safekeeping of materials have yet to be made.

The recommendation concerning the publication of a good and complete book on the history of physics is very important. All the participants at the seminar agreed that such a book, complete, compact, and modern, should be undertaken. Therefore, the seminar proposed the organization of an editorial board which would prepare the book under the supervision of IUPAP and IUHPS. The book would be written by historians of physics (not more than ten authors) and the book would be translated into several languages. Such a book would be of great help to in-service teachers and pre-service student teachers. This would not be a handbook but would provide widely ranging and up-to-date references and would make it possible for the reader to treat special problems in the history of physics in depth. It would consider all aspects thoroughly enough to provide economic and social background material of various historical periods.

A proposal was also made for the editorial board. The editor-in-chief would be Prof. Max Jammer of Israel, and the Commissions on Physics Education of the IUPAP and the IUHPS would recommend members.

The seminar asked the International Commission on Physics Education to cooperate with the Commission on Education of the IUHPS and other scientific associations to set up an international committee for translating materials concerned with the history of science. It would collect recommendations on which books and papers of special historical interest should be translated and would centralize the information on planning translations, including who should be the translators and editors.

The discussions of the seminar showed that the exchange of experiences in the use of the history of physics in physics teaching was very fruitful. On the pedagogical side, up until now there has been only a slight effort to widen the topics of physics and to make the history of physics more understandable. With this sentiment we have begun to make real efforts in this direction; this must be greatly increased.

A. V. Baez (U.S.A.): There were three things that were happening in UNESCO about three years ago which I think may bear some relationship to the work you were reporting. We did an experiment in making documentary films of two scientists. One was Linus Pauling and the other was Louis de Broglie. These films are interviews with the two Nobel prize winners, and we had in mind that we might be able to make a series of films to emphasize not only the technical aspects of the work of Nobel laureates but in one way or another to bring out some of the human implications of what it is that drives men to go into science. These two films have been seen by very few people because it had not been foreseen how they would be distributed. Therefore I am simply bringing to your attention that these two films exist. Surely UNESCO is the proper international channel through which future films might be made and distributed.

Another point is that I remember having seen both at Uppsala and at the Cavendish Laboratories in England extensive displays of scientific material, and I am sure that the continent of Europe

is rich with things that are involved in the history of physics and
of science in general.

S. C. Brown (U.S.A.): In this connection let me read a comment from the Report of the International Working Seminar on the
Role of the History of Physics in Physics Education about information on what, where, and how to preserve historically important material: "For this effort Dr. Charles Weiner, Director of
the Center for the History and Philosophy of Physics at the American Institute of Physics (335 East 45th Street, New York, New
York 10017) has offered the services of the Center to the international physics community. Help would be given in establishing
similar centers in other countries, and where no centers exist,
it would be prepared to advise on where personal papers, photographs, historical apparatus, and other appropriate materials
could be deposited."

The guideline that pointed out the importance of educational
psychology, motivation of students, teacher-student relationships, group dynamics, and the learning processes was well documented in a paper by P. E. Blosser and R. W. Howe (U.S.A.)
under the title: "An Analysis of Research Related to the Education of Secondary School Science Teachers." A few of the highlights of their paper are:

The studies cited and others reviewed tend to indicate that
there are significant similarities and differences in teaching behavior. There is considerable agreement concerning patterns of
teaching behavior; differences and similarities have been identified among various subgroups of teachers. Descriptions of both
verbal and nonverbal teacher behavior indicate that the analyses
of these behaviors should provide useful information for the design of teacher education programs. The studies reviewed provide some indication of teaching strategies used by teachers.
There is not yet, however, enough information to generalize.

Spore[83]* developed a list of 60 competencies related to six
roles of a teacher. The six roles considered were: (1) director

----

*The numbers given in this paper refer to the alphabetized list of
references appended to the paper and reproduced in full beginning
on page 92.

of learning, (2) counselor and guidance worker, (3) mediator of
the culture, (4) link with the community, (5) member of the school
staff, and (6) member of the profession. The 60 competencies
were submitted to four groups of educators to be ranked from
most important to least important roles. These four groups were
administrators, science teachers, science educators, and college
faculty who taught foundations of education courses (philosophy,
psychology, sociology). All four groups ranked the second role,
of counselor and guidance worker, as the most important role of
the teacher. This educational role frequently receives little attention in the pre-service teacher-education program.

Farmer[22] investigated the image of the competent secondary-
school teacher as seen by teachers themselves, administrators,
supervisors, industrial and research scientists, and members
of several national curriculum committee groups. He found what
he considered to be substantial agreement among the groups regarding the competency factors. The most important area of competency was found to be that of the effective use of laboratory work
to teach methods by which scientists have solved problems and to
help students learn to identify problems and solve them empirically. Skillful handling of student questions also received high
priority. Skill in conducting class discussions which stimulate
students to evaluate critically and understand materials more
fully was ranked third in importance of the 16 competencey areas
investigated in Farmer's study.

Teacher competency investigations have provided the rational judgments of a number of persons concerning what they believe
to be abilities which science teachers should possess. While
there were similarities in the competencies reported in the studies analyzed, there were indications of differences in the importance of some competencies and indeed of the functions of the
teacher.

Many investigations have focused on the study of teacher
personality traits, needs, and values: Blankenship,[10] Blankenship
and Hoy,[11] Evans,[21] Kleyensteuber,[37] Lee,[39] Levine,[41] McLeod,[47]
R. M. Merrill,[50] Merrill and Jex,[51] Morris,[54] Navarra and Dugan,[56]
Sargent,[71] Snyder,[82] Walberg,[89] Walberg and Anderson,[90] and Walberg and Welch.[91]

If personality patterns that characterize successful science
teachers could be found, such information could be used in re-

cruitment and selection of teachers. This information could also
be used to guide the development of learning experiences to influence teacher behavior.

Relationships that were identified in several of the studies
are listed:

1.
Several teacher personality traits were identified that related
positively to teaching success and/or to classroom climate.
These included such areas as the teacher's personal adjustment
and personality traits exhibited in the classroom in the form of
pupil-teacher relationships.

2.
The academic preparation of teachers in science related positively to teaching success in several studies. Broad preparation in
the sciences appeared to be more desirable than narrow specialization. The amount and kinds of science experiences teachers
should have in college courses could not be determined from
these investigations.

3.
The procedures used in teaching science classes were related to
teaching success as defined in several of the studies. Several
teaching patterns were found to be related to teaching success,
depending on the method used to analyze the teaching act and the
definition of teaching success. High involvement of students in
learning activities characterized the procedures of individuals
identified in most studies as being successful teachers. The
amount and kind of involvement appeared to be related to classroom climate.

4.
The investigations tended to indicate that teaching procedures are
not equally effective in attaining many different objectives of science education. The teacher's knowledge and acceptance of broad
objectives of science education may be important factors in the
objectives stressed and the procedures used.

5.
The investigations provided data that indicate that the teacher
and the procedures he uses tend to differ in their effect on different groups of students. Scholastic ability, interests, socioeconomic background, student self-concept, and student perception of the classroom were some of the student factors identified
as important learning variables.

A joint committee of the National Association of State Directors of Teacher Education and Certification and of the American Association for the Advancement of Science[55] developed a set of guidelines for the education of secondary-school science and mathematics teachers. These guidelines placed emphasis on the need for depth and breadth in preparation in the sciences. Stressed also were experiences in the methods of teaching the teacher's major science area.

These surveys and others analyzed but not cited emphasize two basic problems: the need for good pre-service programs and the need for strong in-service education programs. Many teachers apparently lack sufficient experience in science to develop an adequate understanding of the concepts and processes of science. Many also appear to have had inadequate education in procedures for guiding students in learning activities.

Microteaching has been used in several teacher education programs to develop a variety of teaching skills. This is a technique in which a teacher teaches for a short segment of time, focusing on specific aspects of the teaching act. Teacher educators at Stanford University[2] have used microteaching extensively in the program for preparing interns. Videotaping of the teacher's behavior allows him to view himself as he appears to others and provides a record to be used as a basis for comparison of his acquisition of skills during the program. Data reported from the Stanford program indicate that there are a number of immediate outcomes, but no research reports were identified that detail the performance of those who have participated in the intern program and are now in their own classrooms.

Goldthwaite[31] conducted an investigation with pre-service teachers to determine whether presenting science demonstrations on a teach-reteach basis would result in immediate improvement in the effectiveness of the teacher in presenting demonstrations. He was also interested in learning whether the students who had participated in the microteaching experiences would present demonstrations more effectively during their student teaching than those who had not participated. He found that those student teachers who had participated as pupils in the microclasses received higher ratings on the effectiveness of their demonstrations than did the students who had been teachers of the microclasses or who had not participated. His data suggest that learning results from participation in such microteaching classes.

Simulation techniques have been used in several universities in the general preparation program for teachers. Only one report specifically involving secondary-school science teachers was identified. Lehman[40] used simulation techniques in an experimental program for secondary-school student teachers of science. The students performed in five basic instructional roles: motivator, critic-evaluator, discussion leader, subject-matter representative, and adapter-modifier. The problem situations focused on basic conflicts in the area of interpersonal student-teacher relationships and dealt with such activities as motivating pupil interest, adapting instruction to differences, questioning, bugeting time and controlling tempo, and problems of pupil control and behavior.

Studies assessing the contribution of other types of instructional procedures for preparing science teachers were not identified. Many articles can be found in the literature in which programs, classes, and experiences for education science teachers are described. Few studies could be found that indicated the kind of teacher produced by a program or the kinds of behavior, competency, skill, or characteristics developed through planned experiences. It would appear that little is known concerning the specific effects of teacher-education programs on science teachers.

A few studies were identified in which general outcomes of teacher education programs were analyzed. Several studies have been conducted to investigate the critical thinking and problem-solving abilities of science teachers: Andersen,[3] Craven,[19] George,[29] Sieber.[79] Data gathered in these investigations tend to indicate that critical thinking, as evaluated, is not a major learning outcome of the study of college science. It is possible, however, that the evaluation instruments used in these studies did not assess elements of critical thinking developed in science courses.

One of the most useful parts of the paper by Blosser and Howe was an extensive bibliography of the United States literature on the subject of educational research related to secondary school teaching. This bibliography is here reproduced in full:

1.
Adalis, Dorothy. "An Appraisal of Broad Subject-Matter Areas

in the Pre-service Preparation Program of Biology Teachers in
West Virginia." University Microfilms, Ann Arbor, Michigan.
1965.
2.
Allen, Dwight. "Micro-Teaching: A Description." Stanford School
of Education, Stanford University, Stanford, California. 1968.
3.
Andersen, Hans O. "An Analysis of a Method for Improving
Problem-Solving Skills Possessed by College Students Preparing to Pursue Science Teaching as a Profession." University
Microfilms, Ann Arbor, Michigan. 1966.
4.
Anderson, Kenneth E. "The Relative Achievements of the Objectives of Secondary School Science in a Representative Sampling of Fifty-six Minnesota Schools." University Microfilms,
Ann Arbor, Michigan. 1949.
5.
Anderson, Norman D. "An Analysis of Programs for the Preparation of Secondary-School Science Teachers." University Microfilms, Ann Arbor, Michigan. 1965.
6.
Balzer, A. L. "An Exploratory Investigation of Verbal and Non-
Verbal Behaviors of BSCS and Non-BSCS Teachers." Unpublished
doctoral dissertation, The Ohio State University, Columbus, Ohio.
1968.
7.
Barnes, L. W. "The Nature and Extent of Laboratory Instruction in Selected Modern High School Biology Classes." University
Microfilms, Ann Arbor, Michigan. 1966.
8.
Biddle, B. J., and Ellena, W. J., editors. Contemporary Research on Teacher Effectiveness. Holt, Rinehart & Winston,
New York. 1964.
9.
Black, H. T. "The Physics Training of Indiana High School Physics Teachers." College Journal 32, 12-14, March 1962.
10.
Blankenship, J. W. "An Analysis of Certain Characteristics of
Biology Teachers in Relation to Their Reactions to the BSCS Biology Program." University Microfilms, Ann Arbor, Michigan.
1964.

11.
Blankenship, J. W., and Hoy, Wayne K. "An Analysis of the Re-
lationship Between Open-Mindedness and Closed-Mindedness and
Capacity for Independent Thought and Action." Presented at a
meeting of the National Association for Research in Science Teach-
ing, Chicago, Illinois, February 1967.
12.
Blum, Sidney. "The Value of Selected Variables in Predicting
Rating Success in Teaching Science." University Microfilms,
Ann Arbor, Michigan. 1966.
13.
Brandwein, Paul F. The Gifted Student as Future Scientist—
The High School Student and His Commitment to Science. Har-
court, Brace & World, Inc., New York. 1955.
14.
Bruce, M. H., McLeod, R. J., and Matthews, Charles C. "A
Study to Identify Relationships Between Behavior Patterns and
Personal Traits in Science Student Teachers." Presented at a
meeting of the National Association for Research in Science Teach-
ing, Chicago, Illinois, February 1967.
15.
Caldwell, Loren T. "Subject Organization Proposed for Teaching
Related Fields Concepts Required to Learn Earth Science Con-
cepts in the Junior High School." Science Education 50, 26-31,
February 1966.
16.
Commission on College Physics, Panel on the Preparation of
Physics Teachers. "Preparing High School Physics Teachers."
Commission on College Physics, College Park, Maryland. 1968.
17.
Connally, G. Gordon. "Multiple Nested Curricula and the N-6
Major." Journal of Geological Education 14, 54-56, April 1966.
18.
Corrigan, Dean, editor. A Study of Teaching. The Association
for Student Teaching, National Education Association, Washing-
ton, D.C. 1967.
19.
Craven, G. F. "Critical Thinking Abilities and Understanding of
Science by Science Teacher-Candidates at Oregon State University." University Microfilms, Ann Arbor, Michigan. 1966.

20.
Davis, C. R. "Selected Teaching-Learning Factors Contributing
to Achievement in Chemistry and Physics." University Micro-
films, Ann Arbor, Michigan. 1964.
21.
Evans, T. P. "An Exploratory Study of the Verbal and Non-Verbal
Behaviors of Biology Teachers and Their Relationship to Selected
Personality Traits." Unpublished doctoral dissertation, The Ohio
State University, Columbus, Ohio. 1968.
22.
Farmer, Walter A. "The Image of the Competent Secondary-
School Science Teacher as Seen by Selected Groups." University
Microfilms, Ann Arbor, Michigan. 1964.
23.
Fawcett, Claude W. "The Skills of Teaching." Students' Store,
University of California, Los Angeles. May 1965.
24.
Ferguson, Max B. "A Curriculum for Training High School Bi-
ology Teachers Which Administrators of Teacher Training Institutions Will Support." The American Biology Teacher 24, 337-
339, May 1962.
25.
Fleek, J. B. "General Chemistry for Prospective Teachers of
Physical Sciences in High School." University Microfilms, Ann
Arbor, Michigan. 1956.
26.
Fullwood, W. E. "Selected Factors as Related to Achievement
in College Physics." University Microfilms, Ann Arbor, Michigan.
1965.
27.
Gallagher, James. "Teacher Variation in Concept Presentation
in BSCS Curriculum Study." Boulder, Colorado. January 1967.
28.
Gallentine, J. L., and Solbert, A. N. "Factors Relating to Success in Teaching Modern High School Biology." Science Education 51, 305-309, April 1967.
29.
George, Kenneth D. "A Comparison of the Critical-Thinking Abilities of Science and Non-Science Majors." Science Education 51,
11-18, February 1967.

30.
Ginsberg, Benson E., et al. "Preparing the Modern Biology Teacher." BioScience 15, 769-772, December 1965.
31.
Goldthwaite, D. T. "A Study of Micro-Teaching in the Pre-service Education of Science Teachers." Unpublished doctoral dissertation, The Ohio State University, Columbus, Ohio, 1968.
32.
"Guidelines for Content of Preservice Professional Education for Secondary School Science Teachers." A Statement by the Association for the Education of Teachers in Science and the Cooperative Committee on the Teaching of Science and Mathematics. The Science Teacher 35, 85-90, May 1968.
33.
Howe, Robert W. "The Relationship of Learning Outcomes to Selected Teacher Factors and Teaching Methods in Tenth Grade Biology Classes in Oregon." University Microfilms, Ann Arbor, Michigan. 1964.
34.
Jackson, Philip. "The Way Teaching Is." Association for Supervision and Curriculum Development and the National Education Association, Washington, D.C. 1966.
35.
Kimball, M. E. "Opinions of Scientists and Science Teachers About Science." University Microfilms, Ann Arbor, Michigan. 1965.
36.
Kleinman, Gladys S. "General Science Teachers' Questions, Pupil and Teacher Behaviors, and Pupils' Understanding of Science." University Microfilms, Ann Arbor, Michigan. 1964.
37.
Kleyensteuber, Carl J. "Group Classifications of Teachers with Evaluative Attitudes Favorable to Science Study." Science Education 45, 236-237, April 1961.
38.
Kochendorfer, Leonard, H. "Rationale of High School Biology Teachers Using Different Curriculum Materials." University Microfilms, Ann Arbor, Michigan. 1966.
39.
Lee, Eugene C. "Career Development of Science Teachers." Journal of Research in Science Teaching 1, 54-63, March 1963.

40.
Lehman, David L. "Simulation in Science Teaching—A Preliminary Report." Presented at the meeting of the American Associa-
tion for the Advancement of Science, Washington, D.C., December 1966.
41.
Levine, Benjamin. "Characteristics of Prospective Science Teachers as Compared with Those of Prospective English Teachers."
University Microfilms, Ann Arbor, Michigan. 1964.
42.
Loud, Norman D. "The Administrator, The Teacher, and The
Students Evaluate a Science Teacher—A Comparative Study."
University Microfilms, Ann Arbor, Michigan. 1964.
43.
Lozanoff, Paul. "The Effectiveness of Higher Institutions in the
Preparation of Biology Teachers." The American Biology Teacher 27, 18-19, January 1965.
44.
Matthews, Charles C. "The Classroom Verbal Behavior of Se-
lected Secondary School Science Student Teachers and Their Cooperating Classroom Teachers." Presented at a meeting of the
National Association for Research in Science Teaching, Chicago,
Ill., February 1966.
45.
Matthews, J. G. "A Study of the Teaching Process in the Class-
room of Selected Science Teachers." Unpublished doctoral dis-
sertation, The Ohio State University, Columbus, Ohio. 1968.
46.
Mayer, Victor J. "Criterion Model of an Earth Science Teacher
Preparation Program." Science Education 51, 290-292, April
1967.
47.
McLeod, Richard J. "A Study to Identify Relationships Between
Science Student Teachers' Behavior Patterns and Those of Their
Cooperating Teachers." Presented at a meeting of the National
Association for Research in Science Teaching, Chicago, Ill.,
February 1968.
48.
McLeod, Richard J. "Changes in the Verbal Patterns of Student
Teachers Who Have Had Training in Interaction Analysis and the

Relationship of These Changes to the Patterns of their Cooperating Teachers." Presented at a meeting of the National Association for Research in Science Teaching, Chicago, Ill., February
1967.
49.
Meinhold, Russell. "An Analysis of the Scores of Science Teachers on a Test of the Methodology of Science." University Microfilms, Ann Arbor, Michigan. 1961.
50.
Merrill, Reed M. "Comparison of Education Students, Successful
Science Teachers, and Educational Administrators on the Edwards
PPS." Journal of Educational Research 54, 38-40, September
1960.
51.
Merrill, Reed M., and Jex, Frank B. "Role Conflict in Successful Science Teachers." Journal of Educational Research 58, 72-
74, October 1964.
52.
Merrill, William M., and Shrum, John. "Planning for Earth Science Teacher Preparation." Journal of Geological Education 14,
23-25, February 1966.
53.
Molchen, Kenneth J. "Verbal Behavior as Related to Local Supervisors as Models." Presented at a meeting of the National Association for Research in Science Teaching, Chicago, Ill., February
1968.
54.
Morris, Kenneth T. "A Comparative Study of Selected Needs,
Values, and Motives of Science and Non-Science Teachers."
University Microfilms, Ann Arbor, Michigan. 1963.
55.
National Association of State Directors of Teacher Education and
Certification. "Guidelines for Preparation Programs of Teachers
of Secondary School Science and Mathematics." American Association for the Advancement of Science, Washington, D.C. September 1961.
56.
Navarra, J. G., and Dugan, R. B. "A Pilot Study to Determine
the Relative Importance of Selected and Professional Factors in
the Success of the Student Teacher in Science." Journal of Experimental Education 31, 413-418, 1963.

57.
Nelson, T. A. "Competencies Desirable for Beginning Science
Teachers as Viewed by Administrators and Science Teacher in
the State of Illinois." University Microfilms, Ann Arbor, Michigan. 1953.
58.
Newton, David. "Research on Science Education Study." Presented at the Association for the Education of Teachers in Science meeting, Washington, D.C. March 1968.
59.
Nicholas, Charles H. "Analysis of the Course Content of the Biological Sciences Curriculum Study as a Basis for Recommendations Concerning Teacher Preparation in Biology." University
Microfilms, Ann Arbor, Michigan. 1965.
60.
Obourn, E. S., Blackwood, P. E., et al. "Research in the Teaching of Science, July 1957-July 1959." U.S. Office of Education,
Washington, D.C. 1962.
61.
Obourn, E. S. and Brown, K. E. "Science and Mathematics Teachers in Public High Schools." U.S. Office of Education, Washington,
D.C. 1963.
62.
Pankratz, Roger S. "Verbal Interaction Patterns in the Classrooms of Selected Science Teachers: Physics." University Mi-
crofilms, Ann Arbor, Michigan. 1966.
63.
Parakh, Jal S. "A Study of Teacher-Pupil Interaction in High
School Biology Classes." University Microfilms, Ann Arbor,
Michigan. 1965.
64.
Reed, Horace, Jr. "Teacher Variables of Warmth, Demand and
Utilization of Intrinsic Motivation Related to Pupil's Science Interests: A Study Illustrating Several Potentials of Variance-Co-
variance." The Journal of Experimental Education 29, 205-229,
March 1961.
65.
Reed, Jack Arthur. "A Description and Evaluation of the Second-
ary-School Science Methods Course in Teacher Preparation Pro-
grams in the United States." University Microfilms, Ann Arbor,
Michigan. 1967.

66.
Regional Science Experience Center Review. Regional Science
Experience Center, Oak Ridge, Tennessee. January 1967.
67.
Rentschler, James Ewing. "A Study of the Academic Training in
Science of the General Science Teachers of Indiana." University
Microfilms, Ann Arbor, Michigan. 1962.
68.
Richardson, John S., and Howe, Robert W. "The Role of Centers
for Science Education in the Production, Demonstration, and Dissemination of Research." U.S. Office of Education, Washington,
D.C. 1966.
69.
Ryans, D. G. "Characteristics of Teachers: Their Description,
Comparison, and Appraisal." The American Council on Education, Washington, D.C. 1960.
70.
Sandefur, J. T. An Experimental Study of Professional Education
for Secondary Teachers. Kansas State Teachers College, Em-
poria, Kansas. 1967.
71.
Sargent, E. A. "A Study to Determine Certain Characteristics
of Earth Science Project Teachers and Students in the Permis-
sive or Authoritarian Classroom Which Lead to Greater Academic Achievement in These Students." Unpublished doctoral dissertation, Colorado State College, Greeley, Colorado. 1966.
72.
Schirner, S. "A Comparison of Student Outcomes in Various
Earth Science Courses Taught by 17 Iowa Teachers." Presented
at a meeting of the National Association for Research in Science
Teaching, Chicago, Ill., February 1968.
73.
Schueler, Herbert, Lesser, Gerald S., and Dobbins, Allen L.
"Teacher Education and the New Media." The American Associ-
ation of Colleges for Teacher Education, Washington, D.C. 1967.
74.
Seeling, R. H. "A Study of the College Science Preparation of
Idaho High School Biology Teachers for the 1964-1965 School
Year." Unpublished Master of Arts in Education thesis, Idaho
State University, Pocatello, Idaho. 1965.

75.
Shannon, John R. "Elements of Excellence in Teaching." Educational Administration and Supervision 27, 168-176, March 1941.
76.
Sheppard, Moses M. "The Relation of Various Teacher and Environmental Factors to Selected Learning of Ninth-Grade Science Pupils." University Microfilms, Ann Arbor, Michigan. 1966.
77.
Shrum, John W. "A Proposed Curriculum for the Preparation of Earth-Science Teachers." University Microfilms, Ann Arbor, Michigan. 1963.
78.
Shrum, John W. "Recommendations for a Basic Academic Preparation for Earth Science Teachers." Journal of Geological Education 14, 26-28, February 1966.
79.
Sieber, Joan E. "Problem Solving Behavior of Teachers as a Function of Conceptual Structure." Journal of Research in Science Teaching 2, 64-68, March 1964.
80.
Skinner, Ray, Jr., and Davis, O. L., Jr. "Preparation of Earth Science Teachers in Ohio." Journal of Geological Education 13, 85-87, June 1965.
81.
Snider, Ray M. "A Project to Study the Nature of Physics Teaching Using the Flanders Method of Interaction Analysis." University Microfilms, Ann Arbor, Michigan. 1966.
82.
Snyder, Norman C. "Socio-Cultural Background Differences in the Personnel of the Social, Biological, and Physical Sciences — A Chapter in the Institutionalization of the Sciences." University Microfilms, Ann Arbor, Michigan. 1965.
83.
Spore, Leroy. "The Competencies of Secondary Science Teachers." Science Education 46, 319-334, October 1962.
84.
Steidle, Walter E. "The Preparation, Certification, and Teaching Employment of Graduates of Science Education Programs in Ohio." University Microfilms, Ann Arbor, Michigan. 1964.

85.
Taylor, T. W. "A Study to Determine the Relationships Between
Growth in Interest and Achievement of High School Science Students and Science Teacher Attitudes, Preparation, and Experi-
ence." University Microfilms, Ann Arbor, Michigan. 1957.
86.
Tubbs, F. B. "Some Characteristics of Highly Effective and Less
Effective Secondary-School Science Teachers." University Micro-
films, Ann Arbor, Michigan. 1963.
87.
Van Allenstein, Richard. "Worth and Technical Competence of
Teachers and the Concomitant Variation in Teacher Effective-
ness." University Microfilms, Ann Arbor, Michigan. 1966.
88.
Van Houten, William. "Laboratory and Demonstration Skills
Needed to Teach Physics and Chemistry in the Secondary Schools —
A Checklist and Suggested System for the Development of These
Skills." University Microfilms, Ann Arbor, Michigan. 1965.
89.
Walberg, H. J. "Personality, Role Conflict, and Self-Concept
in Beginning Teachers." Educational Testing Service, Princeton,
New Jersey. 1966.
90.
Walberg, H. J. and Anderson, G. J. "Classroom Climate and
Individual Learning." Mimeographed. Harvard University,
Cambridge, Massachusetts. 1967.
91.
Walberg, H. J., and Welch, W. W. "Personality Characteris-
tics of Innovative Physics Teachers." Journal of Creative Behavior 1, 163-170, Spring 1967.
92.
Webber, Clemmie Embly. "A Study of the Pre-Service Education
of Junior High School Science Teachers in the South Atlantic
States." University Microfilms, Ann Arbor, Michigan. 1966.
93.
Woellner, E. H., and Wood, M. A. Requirements for Certification. Third edition. University of Chicago Press, Chicago. 1968.
94.
Yager, R. E. "Teacher Effects Upon the Outcomes of Science
Instruction." Journal of Research in Science Teaching 4, 236-
242, December 1966.

95.
Youkstetter, Frank O. "Proposals for the Professionalization of a General Science Course for Prospective Junior High School Science Teachers." Science Education 45, 348-353, October 1961.

The lack of attention to the teaching of applied physics by the conference was protested by J. Macfarlane (Rwanda) in the following statement:

The conference did not discuss the special problems of preparation for the teaching of technical physics or applied schools at the secondary level. Many countries have technical schools at the secondary level; others at an immediate postsecondary level; in these schools physics is taught as a basic element of technological preparation.

The curriculum must be different from that used for the university preparation course, for many reasons:

1.
The objectives of the teaching are different: to provide a sound quantitative basis in physical phenomena basic to common technical applications: mechanics, heat, electricity.

2.
The aptitudes and interests of the students are different. The structure of society is such that technical schools are a second and third choice in a student's orientation.

3.
The time allotted to physics teaching may by necessity be less than in other curricula, since the learning of specific skills is to be stressed.

The methodology used in teaching technical physics must presumably be different, since it must be adapted to the curriculum. In this context, we must note that curriculum development in the field of applied physics has lagged behind that of the more academic course. We note also that teachers of technical physics are most often given no special technological orientation, since they come from the same teacher training institutions.

The curricula of preservice preparation institutions must consider the special problems of teaching technically oriented physics: those of inspiring interest in poorly motivated students,

those of identifying clearly those aspects of physics most pertinent to the technical student.

Chapter 5

In-Service Education of Teachers

An invited paper on "The In-Service Education of Physics Teachers" from E. J. Wenham (U.K.) sets the scene for the report of the working group which concerned itself with this matter:

It is no part of my task to discuss pre-service training; but no consideration of in-service training can ignore what goes before. In my country we prepare our teachers for their life's work by providing either a three-year course in a college of education, a four-year course in a college of education, or a one-year course following graduation within a university. All three routes ensure an adequate background in the material of the student's specialism and in those studies thought to be of importance for the future teacher. In all cases, practical experience is provided, but this is necessarily limited and selective. Large assumptions (which I suspect to be untenable) are made about the transfer of this training to the school situation as the student finds it when he takes up his first post.

Having completed his course, the young teacher enters a school and is there immediately exposed to a process of retraining. It is unfortunately rare for this second process to be related in any way to the first, from which it differs widely. This new process is largely unrecognized, unofficial, uncontrollable, for it is in the hands of each individual new teacher and is directed by the climate of opinion that exists in the school. It arises because the new teacher is presented with a large and continuing number of practical situations, not with the selected few used in his training. It arises because the new teacher is alone with very little chance to discuss his problems with others. In these circumstances, the solutions adopted are largely the traditional ones, and so the traditional style of teaching a subject is perpetuated in spite of all the best efforts of the pre-service trainers. This traditional style is an unchanging skill that assumes that the role of the teacher is unchanging and that he is operating in a static world. In a word, it is a training in conformity. And it is very difficult to resist. On the one hand there is the vast collective experience of a mature staff; on the other, the immature, untried, lonely new

entrant unsupported by contact with the pre-service training institution from which he has so recently come. Small wonder that
the majority do conform and that they abandon that self-critical
appraisal of their own teaching, which the initial training tried
to inculcate, for a static, routine skill that provides clear-cut
answers to well-known teaching problems. That this school retraining is highly successful cannot be doubted; to a considerable
extent it nullifies the pre-service training so expensively provided by the state.

I cannot leave my description of this conflict without indi-
cating where a solution might be found. It is essential that we
provide for better communication between the two phases; staffs
responsible for the pre-service training have both theoretical expertise and successful experience; the teachers in the schools
who are concerned (often without realizing it) in the retraining
of new entrants have constant day-to-day practical experience on
the job. Some means must be found to bring these two skillful
groups of people together.

In a way, this hidden retraining is a part of in-service training, although it would be to everyone's benefit if it were to be
seen as part of the initial training process. As things are, we
must recognize that it exists and that its direction leads us back
to a long-vanished static, unchanging world.

But the situation in the world about us and in the schools
which form a part of that world is neither static nor unchanging.
Indeed, perhaps the major characteristic of our time is the rate
of change in all spheres of life. The situation as it exists demands that our teachers themselves be flexible, self-critical,
prepared for and willing to change. Their task is a complex one,
far more complex than the simple practical skill fostered by so
much of the school retraining. Teacher education must be education for change. Speaking generally, the pre-service courses
provide for this. The in-service courses must restore this element in our teachers where the retraining process has removed
it.

The need for in-service work is now recognized by teachers
and administrators alike. The most usual provision takes the
form of in-service courses. These come in all sorts and sizes—
one-day courses, one-week courses, part-time courses, day-
release courses, full-time courses, vacation courses, one-term

courses, one-year courses, summer institutes, teachers' workshops, and the rest. The assumption is that such courses will
successfully re-equip the teacher and send him back to his classroom remade. So far as my reading of the literature allows, this
assumption is untested. But the empirical evidence seems to be
in its favor.

Let me now turn to the special case of the physics teacher
and concentrate my attention on the in-service phase, leaving the
problem of the retraining to one side. It happens. And in so many
countries we must recognize thankfully that the work of curriculum development groups have made it much more difficult for the
school retraining phase to become so rigid that later change becomes impossible. The work of such groups as the PSSC, Harvard
Project Physics, and Nuffield[1] has, at the very least, made the
idea of a changing physics curriculum respectable.

First I wish to consider certain areas with which in-service
training may profitably concern itself. The first such area has
long been recognized, perhaps to the detriment of equally important matters. It concerns content: the physics teacher's own
knowledge and understanding of that knowledge.

Traditionally this need is met by the short "refresher" course
staffed by physicists from university departments and, rarely,
from industrial laboratories. This provides that splendid "contact with the frontiers of knowledge" which is said to be so necessary for the teacher of physics. It may well be necessary, but
is it possible to achieve? Even if the answer to my question is,
"Yes, it is possible to achieve," we must continue to ask about
the usefulness of the typical course. The problem is a well-known
one; the expertise of the lecturer is usually confined to his scientific specialty, and it is rare to find one who is also a good
communicator. But this is vital. I well remember taking part
some six or more years ago in such a course on solid-state electronic devices organized for physics teachers by a highly respected
polytechnic school and staffed by members of a brilliant research
team. Apart from the first lecture, which was by the leader of
the team (a man with a gift for communicating difficult ideas to
the layman), the course was pure frustration. The lesson was
clear. The selection of lecturers for such courses is a very dif-
----
1. See Appendix C.

ficult business, and one further lesson which can be learned is
that the planning group for such courses must include a schoolteacher who can (and who must) ensure that good communication
develops between the lecturer and his audience.

Such a man can also ensure that the work presented is relevant as well as significant to the teachers attending. Such a man
could have prevented the planners of a course on radioactive-
isotope techniques from teaching 30 physics teachers how to use
1-millicurie sources when the English schools are limited to the
use of 10-microcurie sources.

We need to discover and to use those physicists who are
known to be good communicators to audiences of teachers and
who will recognize that those teachers are very much farther
from the frontiers than most of us would like to believe or are
willing to admit. Once found, such men can provide a unique
stimulation to the teachers and encourage them to take their new-
found enthusiasm back to the schools.

I shall say little more about this level of course. It has a
purpose which is clear-cut, that of upgrading the teacher's own
understanding and knowledge of physics, and it is a style of
course which teachers have come to expect.

There is a second and, to my mind, more important area
with which in-service training must concern itself. This is the
area that includes methodology. Oddly, it is this area that so
often reveals the deficiencies in a teacher's understanding of his
subject. It will include a consideration of the findings of research
into learning, into child development, and so on. It will consider
techniques of evaluation (of progress of the learner and of the ef-
ficacy of the teacher), curriculum development, and others. Once
again, teachers must be involved in the courses from the planning
stage onward. They should be thoroughly experienced, well-trained
teachers who have concerned themselves deeply with the pedagogy
of their subject. Unfortunately (in one sense) such men often grav-
itate away from the school laboratory into those institutions that
are concerned with the initial training of teachers. So, once again,
the planning team for these courses should include that important
person—the teacher who is himself actively engaged in teaching
the age range under discussion. The courses themselves will be
staffed by tutors with a good knowledge of educational research
into the relevant fields as well as thorough understanding of the

problems of learning physics—problems that are perhaps more
complex than those of teaching physics.

Ideally both areas (content and method) should be covered in
all in-service work. The stimulation provided by a good session
on a new topic in physics should evoke the response: How can I
get some of this stimulation into my own teaching?

There exist other important functions for in-service work.
For example, let me remind you of the loneliness of the average
physics teacher. It is most likely that he is the only of his kind
in the school, not for him the catharsis of confessing how difficult he has found the teaching of such a concept as that of emf to
the children he has just left. The problem is a very real one and
deserves recognition within our in-service work which can then
provide opportunities for physics teachers to confer together
about their work. Let me quote an experience in this field, as
described by Dr. Glassman of the Israel Science Teaching Centre:

"One of the serious problems of the teaching profession is
isolation. Specially, the teacher must analyze and evaluate and
organize the material he is to teach, say a chapter in a textbook.
He does not have an opportunity for discussion with his colleagues
on the matter. He does his work in isolation. He then prepares
lesson plans around the material he is to teach. Again, there is
no opportunity for discussion of his work with his colleagues and
he must do this, too, in isolation. An occasional visit by an administrator or an inspector which may or may not be followed by
a critical review of the lesson taught, is of no real practical value to the teacher for reasons so obvious that there is no need to
mention them here. Rarely, and then only in bits and snatches,
do teachers have the opportunity to discuss, in any meaningful
way, with other teachers what they do in the classroom.

With this in mind, we at the Israel Science Teaching Centre
this year conducted a special two-week summer course for vet-
eran teachers who were about to begin teaching the first year of
our three-year sequence in biology. There were a number of lecturers on subject matter, an important lecture on techniques of
evaluation and testing, there were laboratory exercise each day,
but I want to tell you in some detail about the significant didactic
retraining that was central to the course. At the first meeting of
the class, two chapters of the text were assigned for each day.

The class was asked to analyze the chapters for significant material and to divide the chapters into daily lessons. Three members each day were assigned to prepare analyses for submission to the class for discussion and criticism. These three members were asked, too, to prepare a detailed lesson plan for one lesson only from the chapter to be presented before the class. They were asked to consult with each other only as to the subject of the lesson, so as not to duplicate work. Two periods in the morning were devoted to presentation and discussion of the analyses of chapters and two periods in the afternoon were spent on presentation of the lessons, followed by discussion and critique by the class.

Well, the initial response can be imagined. Here were people who had been working for years, doing—for years—what we were asking them to do now. But—they had been working in splendid isolation. Suddenly they were being asked to expose their professional selves, their activities and techniques to—peers! Good Lord! Professional colleagues! Impossible! Psychologically they were devastated. With much hard work and gentle persuasion, with prodding and pulling, we got them through the first few days. Then it was clear sailing, more or less, and it ended with expressions of—well, not appreciation so much as recognition that an important, valuable and eminently relevant professional experience had been undergone. We are convinced that this kind of experience as a way of introducing new ideas and techniques in curriculum and methodology deserves further trial and examination. We are already considering a number of modifications, additions, and deletions, and invite enquiries by anyone who sees some merit in the method.

By way of conclusion let me say that with all the proposed and adopted improvements in content, curriculum and methodology, insufficient time or thought has been spent on the problems of training or retraining teachers for the new programs. No matter how successful a pilot project is reported to be, no matter how positive the research indicates a program to be for the student if the training or retraining of the average teacher cannot be accomplished in a reasonably short time for a reasonable majority of the teachers involved or available, that program will not succeed, and the generation of children may well be lost.[2]

----

2. F. S. Glassman. "Problems of Training Science Teachers." The Junior Science Conference Report. Rehovot, Israel, 1969.

I have already indicated that the traditional standard recipe
for in-service work is the refresher course offered by some institution of higher education. In the U.K. such courses (which
are often planned by the consultants of the Ministry) last for periods of about a week and are residential. In the United States,
the summer institutes are rather longer but not dissimilar ani-
mals. The residential nature of such courses is not accidental—
this is the factor that provides the relaxed atmosphere in which
the teacher participants can discuss their own problems with one
another or with course tutors whose experience is usually wider
than average. In the case of the U.K. courses, I have the impression that they are too concentrated: that too many hours of
each day are filled and that intellectual indigestion may follow.

Other patterns include the meetings of the physics teachers'
centers that have been set up in many British universities with
the active help and encouragement of the Institute of Physics and
Physical Society (IPPS). I have no doubt as to the value of the
work done by these centers: And I have every regard for the work
of the IPPS in encouraging the universities to adopt the idea.

Elsewhere in Britain, teachers' centers are being established
by local education authorities. For example, both London and
Birmingham have splendidly equipped science teachers' centers
that serve as centers of in-service training in the science within
their own areas.

My own experience arises from a type of course that has been
developed with some success in the U.K. since the spate of curriculum development struck the schools. You will know that the
British Nuffield Foundation launched an ambitious development
program that was to concern the teaching of the sciences at all
levels from the infant school to the undergraduate class. The intention was to revitalize science teaching; to make it relevant,
significant, enjoyable, and in tune with contemporary science.
From the outset it was evident that the new materials that were
to be offered to the teachers would demand retraining if the best
use was to be made of them. Such a mammoth task could not be
handled by the Foundation, but only by the local education authorities that are responsible for the schools. If we consider the case
of the so-called "O" level physics program (a five-year course
in physics for 11- to 16-year-old pupils drawn from the top 30
percent of the whole school population and terminating in the "Or-

dinary" level examination for the General Certificate of Education), some hundreds of short courses have been provided in all parts of the country. And the need for such courses is a continuing one. The problem which follows is that of staffing the courses. At first, this was done by members of the project team to whom we soon added teachers of the program in the trial schools (that is, the schools in which the experimental teaching was carried out). But as the courses develop, so more and more tutors are needed to staff them.

Quite early in this process, the Department of Education and Science (the Ministry) established a very small number of 10-week-long courses which take place during school terms so that the teachers attending have to be freed by their schools. This in itself presents grave difficulties which are perhaps specific to the situation in th U.K. and which it is not necessary to discuss. Instead, let me consider the experience of my colleagues and myself in running one of these courses at the Worcester College of Education for a matter of several years. We have mounted 14 such courses in all, only 72 teachers have attended—an average of 5, with a range from 1 to 11. Of these 72, 23 have come to us from 15 countries other than the U.K.—from countries as far apart as Australia and El Salvador, Japan and Uruguay.

From the outset we were determined that these courses should be flexible in organization, informal in manner, and relaxed in tone. We anticipated that the teachers attending would have differing, even widely differing, interests and needs. We accepted that the flexibility we felt to be vital would make heavy demands upon ourselves, but we felt that the price was worth paying.

Typically our courses start with a fairly full, well-documented program for the first week. The teachers come expecting to be worked hard on traditional lines. It would be a pity to disappoint them. So, in this first week, we provide a formal session each morning, practical work in the laboratories each afternoon, a reading program, and a film program. At the end of the week, when we all know one another a little better, we start to enquire about their views on the future program. We explore their interests and their needs and offer a draft of a program for the second week with a forecast (very tentative) of the third week. This process then continues: planning the program in detail one week in advance and in sketch plan two weeks in advance.

We settle down to a pattern that includes one formal session per day. On four occasions in the week this is usually a tutor-led discussion with experimental demonstrations; on another occasion it is again tutor-led but the discussion is more likely to be concerned with some educational topic such as examinations, evaluation, the learning process, and so on.

Other activities include laboratory work, the viewing of films and loops, seminars on various topics, visits to schools and to the physics department of our local university, to a school-equipment manufacturer, and so on.

We have learned not to be afraid of extending the teachers' own knowledge. As interest in the forthcoming Nuffield advanced level courses has grown, we have found it necessary to include sessions on the material which is new to so many of the teachers and which they need first to study at their own level. We have learned that discussions about educational research and its findings are best handled by ourselves rather than by the experts in education upon whom we had expected to draw (perhaps we are less suspect). We have observed with interest how the teacher participants learn to relax in the first five weeks of the course and how in the sixth or seventh week they realize that only four or three weeks are left in which to do all those things they now feel that they want to do.

We ourselves have learned as much from the teachers as they have from us. And we have few illusions as to how much they have learned from us—indeed we have learned to regard ourselves as little more than irritants stimulating a process of "group therapy." And we provide the right conditions for such therapeutic work: a good library, a well-equipped laboratory that can offer almost all the appropriate school-type physics equipment as well as some rather more sophisticated tools, stimulating but sympathetic staff members. All this is provided in a congenial, residential community of 1000 students attending courses of initial training and up to 100 attending in-service courses of one kind or another.

We do not hold any sort of examination at the end of the course; but we do hold inquests. Of course, these are often self-canceling. One course makes a suggestion for improvement; the next would have preferred our original approach. Nevertheless, plenty of ideas for our future guidance arise. For example:

Please provide a practical course in elementary electronics—we
do.

Please provide a course in workshop techniques with special reference to the simple servicing of equipment—we do.

Please build up a library of teaching programs, transparencies,
film loops, and so on, so that we may assess their worth ourselves—we do.

Please provide us with handouts giving the exact details of the
new techniques you show us (for example, "instant" photography).

Please provide more opportunity for us to meet physics students
who are undergoing initial training in the college.

And, above all, please retain the flexible program you now
provide.

I will freely admit that the numbers attending these courses
have been disappointingly low (largely because of the additional
financial burden imposed by living away from home for ten weeks,
because of the teacher's natural concern for the pupils in his
school, and because of the difficulty of finding some satisfactory
replacement). But my colleagues and I are convinced that our
courses have met the objectives of providing some teachers of
physics with a deeper insight into the problems of teaching the
subject and of providing some few teachers who can act as centers
of infection in their own home areas and take part in local curriculum development and in the in-service work of the teachers
centers.

The participants tell us that one of the major benefits such a
course provides is the opportunity to relax and to plan one's future work in a relaxed way, not subject to the pressures of the
waiting class. In a way, we are providing a few of our teachers
with a much-needed sabbatical term. With a typical teaching load
of about 24 clock hours a week, the ordinary physics teacher in
an English school gets little chance to do just this.

The lessons we have learned suggest to me that the "group
therapy" which our teacher participants undergo when we have a
larger group of from 8 to 12 is so valuable that the technique could
well be extended. Perhaps the idea that the teacher goes to a
course is misguided. Perhaps the courses should go to the teach-

er—either in their own schools or in the teachers' centers. There, in familiar surroundings, in the presence of colleagues and of an irritant in the form of a consultant, we really might manage to reach the average teacher—the men and women who opt out of in-service training as at present provided. Perhaps then we might succeed in modifying that retraining which I castigated at the beginning of this talk. Perhaps then we could be sure that we were preparing our teachers for the era of change.

In discussion, S. Sikjaer (Denmark) outlined a system of study groups that was being developed in Denmark in an attempt to combat the teacher isolation to which E. J. Wenham had referred. Nearly all teachers were members of such groups, each of which was led by an adviser. Himself a teacher, he had to discuss progress in the world of education as well as progress in the world of physics with his group. This demanded special qualities, release from routine teaching duties for a fraction of the week, and special training. It was hoped that the new degree of M.A. offered by the Physics Institute of the Royal Danish School of Educational Studies would provide this.

S. G. Bronevshuk (U.S.S.R.) gave a brief survey of the system of in-service training of teachers in the Soviet Union. In his country there are 80,000 physics teachers, and the problems of their retraining has been very great. To try to meet these problems they have founded many institutions for various specialties, and retraining is compulsory for all secondary-school physics teachers.

R. C. Waddell (U.S.A.) explained that the Illinois State Physics Project had developed a summer institute that closely resembled the one-term course at Worcester. The effects had been equally good, but the teachers on returning to their own schools to face the hard facts of life often weakened. In an attempt to extend the effects of the program, it had been followed by an in-service institute; this meant that these same teachers returned to the center periodically for say a full Saturday each two or three weeks. These sessions were perhaps even more valuable than the concentrated summer program, because they were held during the time when the teachers were attempting to bring about the changes that were desired and were facing the consequent problems on a day-to-day basis. They now had the opportunity to return to the group to talk over the problems with their fellows at the time they were meeting them. This had proved to be exceedingly valuable.

The working group on the In-Service Education of Physics Teachers reported:[3]
        In-service education or training (IST) serves various functions, but its prime function is to help all teachers to "grow" and to become better teachers. This growth has five primary dimensions:
1.
Updating the teacher's knowledge and developing a deeper understanding of physics.
2.
Providing the teacher with a familiarity with new materials.
3.
Ensuring that the teacher is acquainted with and involved in curriculum innovation.
4.
Pedagogy; such relevant aspects of the learning process as goals, classroom techniques, student behavior, evaluation, and so on.
5.
Links with other disciplines and also links with industry. The latter will provide an adequate insight into the role of the physicist in industry and current industrial applications of physics.

        A consideration of these led the group to accept the following points:
1.
The recommendation of the Royal Society[4] that teachers should "attend, on average, a week of in-service training a year" was supported as a minimal requirement. Optional attendance en-

----

3. Members of the working group which discussed this subject were: F. Watson (U.S.A.), Chairman, M. Underwood (U.K.), Rapporteur, G. Cortini (Italy), H. P. Hooymayers (Netherlands), P. Martinot-Lagarde (France), J. A. Rodriguez (Venezuela), E. M. Rogers (U.K.), S. Sikjaer (Denmark), R. C. Waddell (U.S.A.)
4. In-Service Training for Teachers of Mathematics and Science in Schools. The Royal Society, London, 1970.

couraged by incentives was preferred to obligatory attendance. However, it would be useful to have wider dissemination of the experiences gained in those countries where such courses are compulsory (for example Bulgaria, Czechoslovakia, German Democratic Republic, and U.S.S.R.) Teachers following both full- and part-time in-service training (IST) need financial support.
2.
The problems of initiating, organizing, and supporting full and part-time IST are very different. While lengthy full-time courses would be the responsibility of the universities, comparable institutions, teachers' colleges, or specially established IST centers,[5] many part-time arrangements could be made on a local basis, yet often with coordination from outside.
a.
Local IST could be encouraged by the appointment of an adviser, a successful teacher (normally both highly qualified and competent) who would be given reduced teaching commitments. Such posts might help to provide a career structure for distinguished classroom teachers.
b.
One approach to IST is to get teachers to help themselves. Small group (typically 8 to 20 physics teachers) activities[6] were strongly supported, and it was hoped that these could take place in "released time." Links with universities and comparable establishment could be provided through the teacher-advisers.
c.
Single courses are often ineffective. The experience of Illinois[7] suggests that some kind of supporting work with follow-up is essential. Their summer school for physics teachers, followed by twelve Saturday sessions spread over one year, provides a working example.

----

5. Paper contributed by S. Sikjaer. "The Work and Duties of the Physics Institute at the Royal Danish School of Educational Studies."
6. Paper contributed by M. Underwood. "On-Going Curriculum Development and the Problem of Achieving it in the U.K."
7. Paper contributed by R. J. Miller. "The Involvement of Illinois Colleges and Universities in Cooperatively Improving Physics Education."
Paper contributed by M. R. Mayfield. "Physics: The Program for Teachers."

d.
The provision of teachers' centers on a local basis and in universities and comparable institutions should be encouraged.[8]
They provide a further agency for eliminating the isolation of physics teachers.
e.
The effectiveness of other agencies for IST needs to be assessed: correspondence courses, radio, television, teachers's journals, and so on.
3.
The mere provision of funds is not enough to guarantee success. The experience of the National Science Foundation in the U.S.A. is illuminating. While much has been achieved, vast problems remain. The initial efforts supported subject-matter courses for teachers; the importance of the relevant aspects of the learning process was learned by hard experience.

One of the points in this working group's report was not accepted by all the members of the Conference. In considering the problem of in-service training, the group report stated:

OPTIONAL ATTENDANCE ENCOURAGED BY INCENTIVES WAS PREFERRED TO OBLIGATORY ATTENDANCE.

S. G. Bronevshuk (U.S.S.R.): I would hope that the final report would say that to achieve higher teacher qualifications, retraining courses should be compulsory.

8. Paper contributed by O. P. Puri. "In-service Institutes in Physical Science." 1970.
Paper contributed by W. Eppenstein. "Summer-Institutes for Secondary School Physics Teachers." 1970.
The School Council. "Curriculum Development. Teachers groups and Centres." Her Majesty's Stationery Office, London, 1967.
Paper contributed by G. Cortini and G. Segre. "In-service Education of Teachers at the University of Naples." 1970.
Paper contributed by E. J. Wenham. "The In-service Education of Physics Teachers." 1970.
Paper contributed by D. A. Tawney. "The Role of Physics Centres in Initial and In-service Training." 1970.
Paper contributed by W. Johnson, "The Role of Regional Physics Associations in the U.S." 1970.

During the past decade the Royal Society (U.K.) has taken a
very active part in stimulating interest in the training of mathematics and science teachers in the United Kingdom. Several reports have been published, notably "Teacher Training for Physics Graduates; a Commentary," 1969; "Teacher Training for Science and Mathematics Graduates," 1970. In a recently prepared
report of the Society it was recommended that

1.
Science and mathematics should be given top priority in the national effort to provide in-service training for schoolteachers.
The case for this priority rests on the increasing need to give
all schoolchildren better opportunities to acquire a good understanding of science and mathematics, the acute and probably
growing shortage of teachers qualified in these subjects, and the
present rapid developments in the content and concept of these
subjects.
2.
Determined efforts should be made to ensure that in the next six
years the vast majority of teachers of science and mathematics
attend, on average, at least a week of in-service training a year.
3.
Local education authorities that do not already do so should be
urged to provide the necessary leave, maintenance, and travel
allowances to teachers attending approved courses.
4.
There should be special publicity drives, coupled with systematic persuasion by inspectors and others in authority, area by
area throughout the country, to encourage attendance at courses.
5.
Salary additions for successful completion of approved courses
of in-service training should be introduced.
6.
Continued and increased support should be given to teachers'
centers of various kinds, and steps should be taken to avoid possible confusion or duplication of efforts without in any way im-
pairing local and subject-based enthusiasms and voluntary efforts.

It is of interest to examine a wide selection of the techniques
adopted in various countries in an attempt to provide a satisfac-

tory structure for in-service work. Many papers were contributed
which indicate how this is being done and how it might be done.
    An unusual approach has been taken in Denmark, where a
physics institute has been established at the Royal Danish School
of Educational Studies, an old, established institution with university status. S. Sikjaer (Denmark) reports in his paper "The
Work and Duties of the Physics Institute of the Royal Danish
School of Educational Studies":

    Although the main task is still to give the traditional in-
service training, we now have the additional obligation to teach
faculty members of teachers' training colleges and to offer not
only elementary one-year courses but also more advanced courses
which presently extend through studies leading to a master's degree. In physics we still have quite elementary courses intended
for strengthening the background of poorly prepared physics teachers. We have courses in woodworking and in metalworking where
the students learn to make and to repair physics teaching apparatus. We have courses where we use calculus and where the students can get a good understanding of classical mechanics, classical electricity with Maxwell's equations in integral form, special relativity, a little thermodynamics, atomic theory, Heisen-
berg's uncertainty principle, and so on. Students from the courses
on this level, where we demand the most, can pass any examination that is both professional and pedagogical.
    Two years ago we started studies for the Master's degree.
A student can start this study only when he has been a teacher
for at least two years, has the knowledge required in mathematics, physics, and chemistry at the level of the final examination
in grammar schools on the mathematics-physics side (the entrance examination to the universities), and is able to read English and German textbooks. The teaching is both theoretical and
practical. Laboratory work includes experiments in connection
with the subjects taught as well school experiments.

**A variety of suggestions were included in a paper by M.
Underwood (U.K.) entitled "On-Going Curriculum Development
and the Problem of Achieving It in the United Kingdom."**

    An interesting development of the last few years has been
the establishment of teachers' centers by more of the local edu-

cation authorities.[9] Such centers contain some or all of the following: recreational facilities, resource center, library, course and conference rooms, and advisory staff. They provide a convenient focal point for refresher courses, and typically serve the needs of about 1000 secondary teachers. Although this development, along with a smaller number of physics centers for teachers appearing in a few establishments of tertiary education, promises to do much, refresher courses and the like are not yet a compulsory part of a teacher's job, as they are in East Germany. Often the teachers who would benefit most from the work of these centers are those who do not participate.

Team teaching and interdisciplinary experiments have been tried as novel approaches to problems of the whole curriculum. But they have been tried on such a small scale and within the existing framework that, as yet, they have had little impact.

One part of the advisory service, acting through local education authorities, has already been mentioned. Others exist at a national level. Fact-finding Royal Commissions report on particular problems from time to time and make suggestions for reform. Their advice is directed at the government and merely carries implications for schools. The nationally organized inspectorate plays an important role, particularly during times of major reform. It assists in courses for teachers and visits schools, but an individual inspector's work will range over such a wide area that the direct effect on schools is minimal.

The influence of national bodies, such as the Schools Council and the National Foundation for Educational Research (NFER), is growing. They initiate or support research and development projects and publish books and pamphlets about these. They have no executive authority over the work in schools, so their effect is often slow and piecemeal. There is certainly a need for wider dissemination of information about new and current research, as a recent report shows.[10]

So far we have considered those parts of our educational system which contribute to the development and improvement of curricula and courses and which help to prepare teachers for their

----

9. The School Council. Curriculum Development. Teacher's Groups and Centres. Her Majesty's Stationery Office, London, 1967.
10. B. Cane and C. Schroeder. The Teacher and Research. National Foundation for Educational Research, 1970.

changing demands. Valuable and essential as these services are, they inevitably lag behind and rely too much on direction from outside the school. Something needs to be built into the system which will encourage and ensure growth from within. An arrangement is needed which will bring together a group of teachers with common interests (in our case the teaching of physics) as a natural part of their work. This group needs to be large enough to promote the spontaneous development of ideas and needs to be provided with the facilities to enable these ideas to flourish.

Such a group would form the basis for reforms as part of an ongoing process. Adaption to change would be seen as part of the successful teacher's job in a much more real way than at present. It has already been pointed out that a school is too small a unit to provide a large enough group of teachers with a common interest. On the other hand, a group based on a local education, authority is too large. A viable group would probably consist of no fewer than eight nor more than twenty persons sharing a common professional interest. A federal group of about five schools could achieve this. It would be the responsibility of the local education authorities to form such groups and to give each group an identity by providing facilities attached to one school to enable teachers to meet and work together.

As soon as a federal group of schools had been formed it would be the responsibility of the physics teachers to meet regularly, say once a week, to discuss their work and their problems. Each teacher would have two professional loyalties: one to the pupils and colleagues of a single school and the immediate community of which it is a part, and the other to a small group of colleagues all concerned with improving the teaching of their own particular specialty. Successful teachers could help to direct the work of others but without being removed from work in the classroom, as tends to happen at present. The immediacy of classroom practice would prevent their suggestions from becoming remote, unrealistic, and unacceptable. Obvious extensions to the work of teachers within a small federal group of schools spring to mind:

a.
A unit of ten physics teachers working in neighboring schools would form an ideal team to develop one aspect of the teaching of their subject or to research into pupil learning and understanding —

for example, the teaching and learning of a difficult concept such as entropy. Local education authorities, teachers' colleges and national bodies could assist in the unification of these efforts. Links with bodies having wider responsibilities would provide natural channels of communication for the exchange of ideas and for the collection and dissemination of information and findings.
b.
Teams of teachers from federal groups of schools could serve a useful function in the training of teachers. A trainee teacher attached to a team would not only catch the excitement of ongoing development but would gain from the collective experience of the team. He could also work in a variety of schools without that wider experience seeming fragmentary.
c.
Various organizations[11] are already calling for regular refresher courses for all teachers. In time these must surely come. When they do, how much more effective they are likely to be if every teacher carries with him to the course developmental work from his team, work that he himself is engaged in. The courses would benefit from a genuine two-way exchange of ideas and information. Not only would the teachers gain in stature, but the colleges organizing such courses would be prevented from becoming ivory towers.

A valuable suggestion by R. David (France) was that active teachers should spend a few days every year in a laboratory of a university or of industry. "Engineers and scientists, recognizing their invaluable and often misunderstood social function would certainly give them a good reception."

L. Skrapits (Hungary) working with a colleague, K. Soós, recently prepared a course for retraining secondary-school teachers in Budapest. The character and organization of this one-year course is the following: there are special lectures, exercises, discussions, and demonstrations at the university; mainly, however, the teachers take regular courses at the university.

The participants of the course take part in a one-day workshop in September, a one-week workshop in January, and again
- -
11. See, for example, the report of the Joint Mathematical Council, 1970.

during the spring holidays and in June. In between, they study by themselves and solve problems. The subject of the course covers the whole of physics, including selected topics in modern physics, and selected demonstrations. Also included are considerations of ideology and methodology.

At the end of the course, the participants receive a qualifying certificate. The qualification depends on the activity shown during the workshops, on the written examinations during the year, on the results of the problems solved during the year, and on an oral examination taken at the end of the year.

The most ambitious program for teacher retraining was reported from the Soviet Union by S. G. Bronevshuk (U.S.S.R.):

It is well known that the human brain has a property to forget things, and so of course teachers working in schools will loose after a time the knowledge that they gained at the pedagogical institutes. So the problem is to recover this knowledge, and for this purpose there are about 180 retraining institutions. At these pedagogical institutions we free the teachers in June, give them their salaries, and require them to attend courses to raise the level of their specialties. This in-service training is compulsory, and every teacher has to attend a course every five years; otherwise they may not be allowed to teach.

Recently we introduced a new program in which we laid down new ideas about how science has been developing recently. We teachers are, by our nature, slightly conservative. We are fond of those methods to which we are accustomed. Therefore the teachers did not welcome the introduction of these new programs. This somewhat handicapped the introduction of the program. To make the teachers able to transmit the new ideas to the pupils, we had to give them the new methods, and therefore during the change to the new method all of our teachers had to attend a one-month course to learn the new program.

The range of in-service activities includes such new ventures as that now operating at the University of Naples which was described by G. Cortini and G. Segre (Italy) in a contributed paper "In-Service Education of Teachers at the University of Naples." This is interdisciplinary in that the Seminario Didattico of the university science faculty includes groups working in several disciplines, thus establishing a relation among the teachers of different disciplines through discussion and confrontation of their respective problems and work methods. It is possible for them, in this way, to find a large area of common interest, which includes psychology, sociology, pedagogic, and statistical techniques.

The activities of the Seminario have consisted of cycles of seminars bound together by some common aim in order to keep alive the interest of audiences formed by teachers of several different disciplines. For instance in 1969/70 we had a cycle of lectures on teaching of physics, mathematics, Italian, and biology, all seen in the light of a common pedagogical and psychological point of view. Then, we had a properly articulated cycle of seminars on structuralism, with the participation of leading Italian scholars of mathematics, psychology, linguistics, physics, chemistry, Italian literature; it was concluded by a final lecture of Jean Piaget.

An alternative, also operating in Italy, at the University of Modena, was described in a paper called "In-Service Education of Physics Teachers at Modena University" by A. Loria (Italy). This voluntary course lasts about eight months and is active on two afternoons a week. The participants all possess the "Laurea" obtained after four years of university study. The objectives of this program are

1.
To remind the participants of the most important ideas and to show these from different and sometimes unexpected viewpoints. Some subjects that are new to many have also been considered, such as the principles of special relativity.
2.
To bring the teachers into touch with modern methods and acquaint them with a wide range of teaching aids and individual laboratory work (for example, from the PSSC, Nuffield, and Harvard Projects). This is regarded as the most important aim of the work.
3.
Some advice on pedagogy and psychology is also given along with a few lectures on the history of physics.

L. Varga (Hungary) prepared the following comments about

the system of retraining the Hungarian secondary-school physics teachers:

The Hungarian National Pedagogical Institute has worked out a retraining plan for physics teachers which lasts for periods of three to four years. The retraining is carried out by the National Pedagogical Institute, by institutions doing teacher education, and at county retraining centers. The system takes several forms:

Practical seminars. Here the aim is to help direct the teaching activity. The seminar lasts for two years, having 150 hours of supervised activity per year. The course is organized by the county retraining centers, and the participants are given reduced duties in their schools. The main activity goes on in practice schools and consists of visiting classroom exercises and demonstration lessons. Some of the themes used for physics teachers include: the psychology of learning, the possibility of making the best use of the physics taught in the eight-year primary school, programmed teaching, methods of investigating results, selected sections on atomic physics, on astronomy, on radiation theory, and also how to teach these subjects.

A one-year course in the methodology of teaching. The aim of this course is to teach new progress in physics, psychology, and pedagogy. The course is organized by teachers' colleges in local areas and requires 150 hours of participation. In the most recent program the following topics were included: the psychology of thinking, testing procedures, and selections from thermodynamics, atomic physics, and radiation theory.

National programs. In these programs any teacher in the whole country can participate. Recently we have had two of these programs: one organized by the National Pedagogical Institute, consisted of one- or two-week summer courses in methodology, in electrical measurements, and in instrumentation; and the other, organized by the L. Eötvös Physical Society and the National Pedagogical Institute, is an annual conference for physics teachers which is usually held during the spring holidays and lasts for four days.

Local programs. These are organized by the educational council of the towns and counties and deal with local and current subjects. They are either held on one school day, or from three to ten days during a school vacation.

Teachers' associations. These associations consist of the
physics teachers of a larger school or of a town. They usually
meet once a month to discuss current problems. A methodological journal, A fizika tanitása, is published by the National Pedagogical Institute and appears six times during the academic year.
It also serves to strengthen the in-service training of physics
teachers.

M. R. Mayfield (U.S.A.), in his paper "Physics: The Program for Teachers," has this to say about the future of the graduates of the program:

Graduates of such teacher education will enter their classrooms well prepared and professionally alive. However, experience has shown that well-prepared, professionally alive new teachers have in the past been unable to maintain themselves for any
extended period of time. One major reason for this near-exponential decline in teachers keeping up to date appears to be the
failure of universitites to provide effective follow-up assistance.
Such assistance, if it is to be meaningful, must meet the needs
of the teachers, not just as viewed by university professors but
more importantly as viewed by the teachers and their administrators. University centers, to provide both demanded and center-
initiated services for teachers, must be developed locally throughout as much of the world as is feasible. Such a system of centers,
cooperating through systematic intercommunication of ideas and
activities could deliver a concentration of intellectual energy sufficient in magnitude to solve and to implement the solution of the
major problems of secondary-school physics education, not merely for the present but as they arise in the future. If there are
enough of us willing and able to develop successful pilot centers,
others will follow, and physics education at all levels will prosper.

The goals and the plans of the Illinois State Physics Project,
as envisaged in "The Involvement of Illinois Colleges and Universities in Cooperatively Improving Secondary Physics Education" were considered by R. J. Miller (U.S.A.)

As the project developed, the goals stated were to increase

interest in physics at the high-school level and to increase enrollment in high-school physics.

Student reaction to physics indicated that most of them thought
physics was only for the most gifted and mathematically adept.
Counselors also characterized physics classes in the same way.
Often the counselors received their viewpoint from physics teachers who complained to the counselors when anyone but the very
best was allowed to enter the physics class. To effect any increase in enrollments meant that this attitude toward physics on
the part of physics teachers, counselors, and students must be
changed.

The plans for improvement centered around a program for
secondary-school physics teachers. The salient features were:
1.
Summer institutes were organized and were followed by in-service institutes as a way of pointing out necessary changes and examples of teaching methods.
2.
The focus of the summer institutes as well as the in-service institutes was upon the phenomena of physics. Teachers have been
taught how to show phenomena, involving their students in class
and laboratory participation.
3.
Each participant had the privilege of inviting staff members from
the in-service institutes to visit his high school. The visitor
taught classes, suggested laboratory equipment, and discussed
physics with students, other teachers, counselors, and administrators.
4.
Because the counselor outlines the course of study for a student,
it is important to secure his cooperation. This has been done
with conferences at the in-service institutes and also at individual conferences in their respective schools.

Unity of approach was obtained at directors' meetings held
at Lake Forest College. Professor Harald Jensen was the instructor for these sessions and presented phenomenological physics as a way of teaching. Three to four days were given to the sessions each year. The directors were convinced that the techniques
of phenomenological teaching as advocated by Professor Jensen

were possible and very desirable as ways of teaching at the high-
school level. In fact, these directors were convinced that they
could modify their own college teaching to use this phenomenological approach.

As suggested previously, each institute emphasized phenomena as a way of presenting physics to the high-school student.
This was a change in viewpoint which required considerable persuasion. Teachers were reluctant to change because, for one
reason, the analytical approach required less preparation. With
five or six classes a day the teachers had little free time to prepare apparatus for the class sessions. The greatest barrier was
the statement: "Teaching physics this way means watering down
the course." The institute instructors in the six- to eight-week
sessions convinced many of the teachers that presenting the phenomena of physics is authentic physics.

In-service institutes were held on alternative Saturdays during the school year at each of the summer institute centers. In
addition, extra institutes were held at other locations. Participants of the summer institute programs were required to enroll
in the in-service institutes; they thought the institutes the most
important contribution of the program. Project coordinators and
directors considered the combination vital to the Illinois Physics
Project. The in-service institutes complemented and completed
the summer institute program. The in-service groups became
centers of identification for physics teachers where information
and apparatus were shared.

An unusual but nevertheless valuable initiative was taken in
the United Kingdom by the physics community itself acting through
the Institute of Physics and Physical Society (IPPS). This is described by D. Tawney (U.K.) in his paper "The Role of Physics
Centres in Initial and In-Service Training." He describes the initiative of the IPPS, as follows:

The Purpose of Physics Centers

In the sixties, a decade notable in the United Kingdom for many
innovations in the field of science teaching, not the least important was the formation of local physics centers by the IPPS.

By the mid-sixties it was apparent that the rapid growth in

school sixth forms (pupils of 12 to 18 years), which had doubled
in size during the previous decade, was not continuing to be
matched by a corresponding growth in the number of physics
students entering university departments;[12] furthermore the concurrent rapid growth in the number of teaching staff in institutions of higher education had created a shortage of suitably qualified candidates for school-physics teaching posts.[13] Simultaneously the interest of the IPPS in physics teaching had increased,
one manifestation of this being the increase in the number of articles on this field in the IPPS Bulletin, until the formation of a
separate journal, Physics Education, in 1966.

In 1965, the Joint Committee for Physics Education formed
by the IPPS and the Royal Society appointed Dr. J. A. Clegg as
Educational Consultant, and it is predominantly as a result of his
efforts that, starting with the center at the Manchester Institute
of Science and Technology in 1966, a number of physics centers
were established at universities throughout the country; the center at Keele, which began in March 1968, was the fifth, and by
the end of 1968 had been followed by five more. By the present
day the number has been doubled.

As Clegg points out, each center has developed a character
of its own, but he sees their purpose as supplementing the work
of the Department of Education and Science and of the university
institutes of education, which provide formal in-service courses
by "keeping teachers of more advanced pupils in contact with
other physicists who are actively engaged in research." Clegg
continues:

"On consideration, it seemed that we might further our own objective by setting up a group of a somewhat different kind; one
which would meet in a university physics department and whose
primary function would be to deal not so much with educational
matters (although teaching techniques in curricula would naturally
be part of their concern) as to discuss basic physics. What we had

----

12. Council for Scientific Policy. Enquiry into the Flow of Candidates in Science and Technology into Higher Education. Her
Majesty's Stationery Office, London, 1968.
13. H. G. Judge, Physics Education 1, 267-270, 1966.

in mind was something that could be described as a workshop
which would be a regular meeting with physics dons of a university and teachers from schools in the surrounding area."[14]

Clegg envisages the aims of physics centers as including the
organization of activities around which the informal contact between the teachers from different institutions could grow, and
the provision of rooms for formal and informal activities; accommodation available should include a lecture theater, a common
room, laboratories, and a workshop.

A center should be organized by an elected committee on
which schoolteachers are in a majority. The Keele Centre is
probably typical in having, in addition to six teachers, one representative each from the university physics department (Chairman), the university education department* (Secretary), the physics department of the North Staffordshire Polytechnic, the local
education-authority science advisers, and Her Majesty's inspectorate. The policy of a center determined by its own committee
sometimes departs from Clegg's original concept; most place
more emphasis on matters directly related to the classroom than
would appear to have been his intention, and furthermore they
welcome physics teachers not merely from the sixth forms (containing academic pupils in the age range 16 to 18) but from the
whole range of age and ability.

Their Use for In-Service Training

Arrangements in the United Kingdom for in-service training are
haphazard, with many bodies providing courses of various kinds.

With the exception of courses leading to advanced diplomas
in education or to higher degrees, there is no financial incentive
for teachers to participate in in-service training. Local educational authorities vary in the extent to which they will pay the
course fees and the consequent travel and subsistence expenses
of teachers and in the extent to which they are able to support

----

14. J. A. Clegg, Physics Bulletin 20, 412-414, 1969.
*Concerned with the initial professional training of graduates at
the University and with higher degrees in education (M.A. and
Ph. D.)

teachers for longer courses. A man who for the purpose of attending a course, is obliged to live away from his family inevitably suffers a financial disadvantage.

For these reasons the proportion of teachers undergoing inservice training is small and the lack of both a well-organized national system of in-service training and of appropriate incentives to teachers to participate in in-service training is currently subject to criticism.[15]

It is against this background that any in-service training function of the physics centers must be viewed. At a meeting of representatives of centers in July 1969, it was apparent that the idea of centers having an in-service training role was viewed with suspicion by some members of university physics departments. The success of centers in providing informal contact between teachers from a wide range of institutions could be jeopardized if schoolteacher members felt that the motive behind the attendance of university members was one of pedagogic do-gooding.[16]

Nevertheless an inspection of the programs of centers, most of which bear a striking resemblance, suggests that, whatever the desired image, there is a strong element of in-service training. Of the 39 meetings held at the Keele Centre since it was founded, 8 could be described as providing knowledge of current developments in physics or technology, while 31 provided knowledge of curriculum developments. The financial support that the Keele Centre derives from the local educational authorities and which new centers are receiving from the Ministry is a recognition of the in-service function of centers.

What prevents the occurrence of the "them and us" attitude feared by university members, as a consequence of emphasis being placed on the in-service role of centers, is that the programs are arranged by a committee on which teachers are in a majority. Furthermore lectures of an in-service nature are often given by the teachers themselves; the most successful activity organized by the Keele Center in the 1969-1970 season was a series of four evenings devoted to the teaching of wave motion; three of these were conducted by teachers, all members of the center and one

----

15. J. Classman, Schools Council Project Technology Bulletin 3, No. 12, 1-5, 1970.
16. D. A. Tawney, Education in Science 8, No. 36, 30-32, 1970.

by graduate students in training for a professional qualification in education. The organization of in-service work by teachers for teachers is an encouraging sign of the growth of professional awareness.

In the future, physics centers are likely to organize more study-group or workshop activities along the lines of the wave-motion series or of the work of a group within the Keele Chemistry Centre which is preparing and testing multiple-choice questions. Members of the center who come from institutions of higher education can contribute professional expertise to many aspects of a center's work, but at a time when the teaching methods used in higher education are being questioned, there is no doubt that they can receive as well as give.

Chapter 6

Curriculum Innovation in Schools

The report from the working group on Curriculum Innovation was as follows:[1]

Why is There a Problem?

A physics curriculum is by definition a method of working which enables a physicist to guide the work of his pupils. It is enough to establish the starting level and the general policy. The physicist himself is able to give direction to the studies and therefore to choose the curriculum and subsequently to develop it as the need arises. This is the ideal method, the one chosen by the very able teacher with a small number of pupils, guiding them freely toward the goal he has set for himself. It is he who gives the pupils their certificates. This is the method used in the final period of university studies, in preparing students for doctoral degrees. There is here no problem of innovation in the curriculum, which takes place automatically.

In secondary education, however, the position is quite different. This is a mass problem, generally concerning hundreds of thousands of pupils. The investment needed for textbooks, laboratory materials, the training of teachers, and so on is considerable, and this raises economic problems. It is also impossible to allow each teacher to be responsible for his own curriculum. The curriculum must normally operate within constraints, particularly where pupils are taking an external examination, the syllabus for which often applies to a whole country. The importance of such examinations must be borne in mind; even in the

----

1. Members of the working group which discussed this subject were: M. Y. Bernard (France), Chairman, F. J. Kedves (Hungary), Rapporteur, J. Cessac (France), S. C. Chen (U.S.A.) E. Danilović (Yugoslavia), A. E. M. El Kashef (U.A.R.), E. W. Hamburger (Brazil), L. Holics (Hungary), Y. Kakiuchi (Japan), R. Petit (France), L. Skrapits (Hungary), P. Targov (Bulgaria), Mrs. I. S. Tawfik (U.A.R.), D. A. Tawney (U.K.), J. Vachek (Czechoslovakia), E. J. Wenham (U.K.).

U.S.A., a country otherwise typical of individual effort, there are general examinations organized on a wide basis.

All of this shows that a curriculum at the secondary-school level is a system profoundly integrated into the economic, social, and political system of the country as a whole. The role of physics is relatively modest in the face of these constraints. Curriculum innovation cannot be discussed in the abstract; a large number of limiting factors, not conditioned by physics, has to be taken into account.

Ongoing Innovation

This is the ideal method. It is possible to envisage, for example, a national commission that examines the curriculum annually and decides on modifications. This procedure is followed in many countries, notably the socialist ones, with the help of commissions appointed by the government. Innovations are necessary for a number of reasons:

1.
Because physics is developing. New concepts are emerging, and new models are constructed. It has to be decided whether or not they should be introduced into the curriculum. If yes, their introduction can lead to upheaval in the remainder of the curriculum; this is a question to which we shall return.

2.
Because of technological progress. This makes it possible to present new concepts with the help of experiments that have been impracticable in the past such as the use of Polaroid photography, and so forth. Above all, however, this requires constant revision of the examples chosen to illustrate physical laws. In the space age it is impossible to ignore the man-made satellite when teaching central forces, or the rocket when teaching the principle of action and reaction. If care is not taken, this could lead to the overloading of the curriculum.

3.
Because other disciplines are developing, and especially because chemistry and biology are going beyond the descriptive stage to the stage of explanatory theoretical models. Science is a unity, and pressure toward integrated science teaching leads inevitably to revisions of the physics curriculum, alongside those in chemistry and biology.

4.
Because the progress that has been made in experimental psychology provides a better understanding of the learning process.
5.
Because the aims of physics teaching change with the evolution of society.

As a result of these various influences, and under the control of an ad hoc commission, it is possible to envisage a relatively slow evolution, modifying syllabi year by year. This method is intelligent and effective; it introduces only those innovations that are compatible with economic resources, the capabilities of the teachers, and such outside constraints as external examinations.

In this as in other fields, effective action can be taken by the associations of teachers in secondary and higher education having the aim, not of defending the interests of the teachers, but of enhancing their position professionally. Such associations, if they are strong, can have considerable influence on the responsible authorities, which can avert difficulties in introducing agreed modifications. In addition, such associations publish reviews that enable every teacher to make his views known on one point or another. All this is very effective.

Radical Innovation; Revolution in the Curriculum

However, continuous innovation can produce serious problems, for social constraints are always present and do not change in phase with the changes in physics. As a result several situations can arise:
1.
The change attempted is one of addition and the result is an enormous syllabus overloaded with material that gives an over-all impression of being archaic. The risk of this is great, for there is always a strong temptation to add new material without modifying the whole. It is obvious that such a situation is unstable; sooner or later a complete change will be necessary.
2.
If there is an attempt to change the method by which a particular concept is developed, a chain reaction is set up that affects a

wide section of the curriculum. Generally, those responsible hesitate to cause such an upheaval and thus unbalanced programs
lacking stability arise.

However, more generally, curriculum development can profit from a change in conditions, notably from a rise in the standard of living which allows the school-leaving age to be raised and
which gives a measure of the development of a country.

An unstable curriculum or a change in conditions demands a
total modification of the curriculum, involving a thorough examination ab initio. PSSC[2] Physics and Harvard Project Physics in
the U.S.A. and the Nuffield Science Teaching Project in the U.K.
are examples of such radical changes.[3]

It is instructive to state the conditions necessary to realize
such a project. First one must have a large group of physics
teachers from secondary schools and from higher education. A
teachers' association (the British Association for Science Education in the case of the Nuffield Project) or a university (M.I.T.[4]
for PSSC or Harvard for Project Physics) can direct operations;
however, a national committee can also do this provided it is representative of all the interests concerned. This is not necessarily
the case with nominated committees. There must be money; this
can be provided by a private foundation or by the state. The costs
of such projects (of the order of a million dollars) must not be
underestimated.

Above all, it is necessary to experiment; this is the aspect
of a project which gives the greatest cause for concern. In practice an educational experiment presupposes that a small fraction
of the pupils follows an experimental curriculum, while the majority continues as before. This is conceivable in a country whose
social structure allows different systems of examination to exist,
as in the U.K. or the U.S.A. In France, the situation is more
complicated, and because of the centralized system it is difficult
to instigate the educational experiments necessary. In other countries, notably the socialist ones, there are special classes in
which experiments can be carried out. However, we must not de-
----
2. Physical Science Study Committee.
3. See Appendix C.
4. Massachusetts Institute of Technology.

ceive ourselves; educational experiments are both delicate and indispensable. Again we face the socioeconomic aspects of the curriculum.

However, it is necessary to disseminate the results of such experiments on the widest possible scale. This presupposes a motivation of the teachers which is easy to achieve if the following conditions are obtained:

1.
The experiment is a success; one must discuss the degree of success: Have we achieved a better understanding of physics, or better examination results, or a greater enthusiasm for physics?

2.
Secondary-school teachers have been closely associated with the experiment. This is vital for the operation of any project. Later these teachers can serve as propagandists and motivate the rest of their colleagues. Any innovation planned without such discussion with teachers by a government authority alone or by a university is difficult to put into practice.

3.
Satisfactory material conditions are arranged; there are textbooks readily available; there are in-service courses for the teachers; there is ample apparatus for practical work. On this latter subject we must refer to the methods of working of the different countries. In the socialist states the government has the responsibility for the manufacture and distribution; in France the government has an equipment center that distributes to the schools all the material it has commissioned from private industry; in the U.S.A. each institution buys its own equipment directly from industry. In the U.K., so far as the Nuffield Project was concerned, the organizers made drawings available and encouraged the manufacturers to make their own versions for sale to the schools.

Once the new curriculum has been established on a large scale, an attempt is necessary to preserve its true character, using the methods in the first section. A curriculum is a living thing; we still have to find out if it has a limited life-span, if it is essential periodically to reorganize it completely, or if a system could be discovered which has an indefinite life. This is a major problem: Must a curriculum having a limited life undergo a complete change from time to time or can it, with continued careful and methodical revision, carry on indefinitely to form the basis of an adequate training?

A Curriculum is Not Transferable From One Country to Another

It is paradoxical to point out that although physics is universal, the physics teaching curricula of different countries differ widely. We find yet again the strict relationship between curriculum and socioeconomic conditions. Several experiments have been carried out in which a curriculum developed in one country has been transferred into another country. The results have been disappointing.

Each country, whether developed or developing, must itself attack the problems that curriculum innovation brings. However, it may draw useful lessons from experiments that have been conducted in other countries. It could be a fundamental role for UNESCO to ensure free communication between different countries by distributing to all science teachers' associations the outcomes of all curriculum experiments in some suitable form.

Recommendations

With regard to curriculum innovation in secondary schools, the working group recommends the following:
1.
Emphasis be placed on the fact that physics is international, dynamic, and not allowing of prejudice.
2.
Each country examine its physics curricula and, if radical change is needed, set up reform groups.
3.
Mechanics be taught as part of the physics course.
4.
Local groups of secondary-school physics teachers be formed to meet regularly and to discuss and promote curriculum innovation.
5.
National journals of physics teaching be established where they do not already exist.
6.
The examination which terminates secondary education (or the

university-entrance examination in some countries) be modified
to match new curricula and made sufficiently flexible to permit
pedagogical experiments without harm to the pupils.
7.
Research be carried out to examine the validity and even the necessity of such examinations in the selection of students best
suited to receive higher education.
8.
Similar research be carried out concerning all school examinations.
9.
Each country develop its own curricula, using materials developed
in other countries only as important sources of inspiration and
reference.

The difficulties referred to in the report of the working group
on Curriculum Innovation concern the problems of designing, testing, implementing, and revising new programs. These problems
are illustrated in the succeeding extracts from contributed and
other papers. Thus, E. M. Rogers (U.K.), describing the development of the Nuffield Physics Project for 11 to 16-year-old
pupils in the U.K. says:[5]

Habit and existing textbooks made formal teaching for memorization easy. Just a new syllabus, incorporating some modern
topics, would soon be used for the same formal teaching. The
Nuffield Foundation realized that teachers would need profuse
teachers' guides that would not only outline a new, more con-
certed scheme of topics but would discuss the teaching in detail,
giving reasons and specimens of teaching treatment, suggesting
things to be said (and things not be said) before an experiment,
and comments teachers might make (or, better, elicit from the
class) after it. In other words, guides that could teach teachers
or offer them some new attitudes and methods.

And elsewhere:

----
5. E. M. Rogers, "The Nuffield Project." Background paper for
the Conference. Reprinted from Physics Today 20, 40, 1967.

We are fully aware of the importance of external examinations at the end of our five-year course .... . In fact, even the
most wonderful teaching program that we could imagine would
be largely spoiled within a few years if it had to be tied to examinations that did not fit its methods or spirit.

Teachers could gain some assurance from teachers' guides
that the suggested teaching was feasible and worthwhile, but all
concerned insisted that there must be new public examinations
for any school that tried the suggested program. Otherwise students' chances in that important test would be jeopardized, with
the undesirable result that the new teaching would be ridiculed by
the mismatch.

At an early stage, we consulted the nine Examining Boards,
whose examinations are taken by students on a nationwide basis.
They agreed to set special alternative examinations, not only
based on the topics of our suggested program but with questions
worded and marked in the spirit of teaching outlined in the teachers' guides. In fact, they accepted our 1000-page set of guides
in lieu of the compact syllabus that had always been required!

With regard to school trials of the material, Rogers writes:

With teachers' guides, apparatus guides to help the setting
up of experiments, and some films made for teachers, we ran
rough trials of section of the course in a few schools with team
leaders watching and helping. The next year, we ran careful trials with full apparatus provided in fifty schools after giving teachers a one-week briefing institute.

Trial teachers reported that putting the suggested program
into action is very hard work for the first time round, but almost
to a man they said, "It is worth it." A few who were moved to
other work insisted that they carried some permanent change in
their teaching with them. On the other hand, those teachers who
felt compelled to compress the course into shorter time with
more efficient coaching are meeting difficulties and finding their
teaching moving back toward traditional form—a mere change to
a different choice of topics is not enough.

Commenting on the report of the working group, Rogers
said:

We have been warned that a new program very easily becomes more and more full of content as the years go by until a
new revolution, slow or fast, is necessary. I should like to suggest that this growth of a new program by the addition of new
material is not universally certain. If the program is reviewed
and revised by those concerned with its inauguration, design,
and running, they will, after some years, cut things out rather
than put things in. A combination of teachers practicing within
the program and program organizers is more likely to make the
program grow smaller and more practicable than longer.

With regard to the export of the scheme to other countries,
Rogers commented:

Since this is a five-year course arranged to fit the apparatus
available in England and Wales, its materials are not likely to be
of direct use for teaching elsewhere. Yet the teachers' guides
offer suggestions and commentary that should make them of considerable value in any library for teachers in other countries.

In the ambitious program developed in the state of Illinois
(U.S.A.) some of the typical difficulties of curriculum innovators were met. These are described in a paper by R. C. Waddell,
entitled "The Program of Involvement of Eastern Illinois University." Among several general observations, Waddell notes:
1.
The transfer of textbook knowledge to the laboratory or to the
real world is much more difficult than is commonly recognized.
2.
The teacher is professionally isolated in his school with no one
to turn to for needed help, yet it is assumed by his colleagues
and the public that he is an authority on physics.
3.
The public-school teaching load, which usually allows one one
"free" period per day, does not permit the design of demonstrations and experiments, the repair of equipment, the setting up
of laboratory.
4.
Physics is often taught directly from the textbook, with excessive emphasis on algebraic and trigonometric manipulations.
This approach requires little advance preparation.

5.
Physics is regarded by students, administrators, and the public
as difficult, not relevant to the needs of society, to be taken only
by a select few; yet it is still prestigious.
6.
Often physics classes enroll mainly the gifted students, but
grades given are only average. This discourages students who
need high grade-points.

As the breadth of the problem was recognized, several approaches were simultaneously taken to reduce the difficulties:
1.
Increased emphasis was placed on the teachers relearning physics at their own operational level. There is a tendency for the
teachers in the laboratory to assume they have sufficient knowledge to do, for example, a given PSSC experiment. After a few
minutes of casual manipulation of the equipment they often put it
aside as being understood. In reality they have only a superficial
understanding of the physical concepts, laboratory techniques,
and pedagogical approaches that are necessary. Efforts were
made to have each teacher go through a complete analysis of
given experiments. After a few weeks the significance of this approach was grasped by most of the teachers.
2.
The participants were encouraged to construct and repair their
own equipment. This was the first opportunity for many to get
this close to the experiments or demonstrations that they should
be doing.
3.
A physics lending library of paperbacks, manuals, and textbooks
was established at Eastern University with such materials being
made readily accessible to all area teachers. In addition, specialized equipment belonging to Eastern's physics department
was made available on loan to the teachers. Many of the area
teachers now regard Eastern as a resource center.
4.
Participation in the program resulted in developing friendships
between nearby teachers. Their sense of isolation is being reduced.
5.
During the methodology sessions emphasis was placed on the

participants' responsibility to regard themselves as physics teachers as well as teachers of their own subject of specialization.
6.
A major effort was made by example as well as verbally to emphasize a phenomenological approach to the introduction of concepts. Direct involvement with physical phenomena enhances student interest, reduces student dependence on the teacher, and results in a more rapid grasp of basic principles. Equipment, experiments, and demonstrations developed in many of the major physics-teaching programs were used.
7.
Mathematics is regarded as a desirable tool to be used in describing physical systems but not as an end in itself. High-school physics can be successfully taught using only the concepts of introductory algebra. However, the high-school course can and should use to the maximum the individual students' understanding of mathematics.
8.
The teachers were urged to reexamine their grading policies and perhaps raise physics grades to the level given in most other subjects to students for comparable performance.
9.
The teachers were encouraged to broaden the base of their classes by accepting students from the upper 50 percent of the total student body, to eliminate prerequisites of trigonometry and senior math, and to foster the enrollment of girls in physics classes.
10.
To bring about greater enrollment it was suggested that teachers publicize physics by hall displays, projects, student contacts at the lower grade levels, and through discussions with guidance counselors and principals. Principals and guidance counselors were invited to attend one Saturday session at Eastern to discuss mutual problems.
11.
A staff member acted as a consultant to the teachers by visiting the schools of most participants at the latters' invitation. He talked to groups of interested students, conducted classes, suggested improvements, aided in the laboratories, and consulted with principals and guidance directors. His observations provided a valuable source of feedback to the Institute.

Considering this course in the state of Illinois, R. J. Miller
in his paper, "The Illinois State Physics Project. The Involvement of Illinois Colleges and Universities in Cooperatively Improving Secondary Physics Education," provided guidelines for
widening the appeal of the physics course in a situation where
enrollments in such courses were falling off sharply.

We believe that a first course in physics can be taught so
that it will have a broader appeal to more students. In order for
this to be true, we believe that
1.
Physics must not be made more difficult than other academic
courses.
2.
Students must be able to make good grades in physics, at least
as good as those obtained in other courses.
3.
Girls must be convinced, on the basis of positive experiences,
that physics is not for boys only.
4.
It must be conceded that a first course in physics cannot cover
all topics in physics, and furthermore that topics can be eliminated without affecting the respectability of the course. Even
well-qualified teachers are not conversant with all aspects of
the discipline.
5.
It must be agreed that, while the language of mathematics is a
delightful and useful means of describing the concepts of physics,
a first course in physics might actually serve its purpose as well
or better if direct observation of physical phenomena is made
central to the course. Observation of phenomena followed by appropriate descriptions, mathematical or not, is a more direct
route to understanding for a person making his first contact with
physics.
6.
A course centered around a set of related student activities has
more appeal than one that attempts to be rigorously analytical.
7.
The aspects of physics that are meaningful to youngsters having
one background may not be meaningful to other youngsters.

8.
The teacher must feel completely free to tailor his course in his
school to the unique capacities, abilities, and needs of his students. This must be true even if the resulting course and its
presentation bear little resemblance to what is now considered
a standard physics course.
9.
The teacher must feel that his task provides an intellectual challenge and that creative effort is required to make physics a vital
and relevant part of his students' experience.

**Members of the group during their discussions presented
time scales of secondary-school physics teaching in their countries. In general the timing was not very different; as an example
L. Holics and F. J. Kedves (Hungary) presented this brief summary of the physics curricula used in Hungary for general pri-**
mary and secondary (gimnázium) schools:

Physics education starts in the sixth year (age 11 to 12 years)
of the eight-year schools with 2 contact hours per week; during
the three years this amounts to about 200 hours. In the sixth class
they have phenomenological treatments of the special properties
of materials; forces (changing the shape of a body); density; concepts of heat; and geometrical optics. In the next year the pupils
get some quantitative knowledge of the same topics. In the eighth
class they learn basic electricity.

In the four-year general secondary schools (age 14 to 18),
physics teaching starts in the second class and lasts for three
years, having 3, 3, and 4 class lessons per week (about 300 hours
during the whole course). In the second class there is statics,
kinematics, and kinetics; in the third class, mechanics of plas-
tic solids, periodic motion, optics, and thermodynamics. In the
fourth class there is electricity, radiations, atomic and nuclear
physics, and astronomy.

For the classes specializing in physics, the physics curriculum is the following: 4, 6, 6 class hours per week in the last
three years, which means about 500 hours altogether. During
all of the four years, 2 hours per week is spent in practical work
(shop practice and laboratory); this amounts to 250 hours during
the whole course. The main topics covered are the same as in

the general physics course. During the first year the students learn some general methods of measurement.

At the end of secondary school all pupils have to take a comprehensive examination. In the general secondary schools the topics are Hungarian language and literature, history, and mathematics; it is compulsory to choose one of the following subjects: one foreign language, physics, chemistry or biology. At the end of the special physics course the subjects of the final examinations are Hungarian language and literature, history, mathematics, physics; and the elective subjects foreign language, chemistry, and biology.

The final examination does not allow entrance to university study; there is another special entrance examination consisting of two subjects, depending on the specialties studied.

On several occasions, S. G. Bronevshuk (U.S.S.R.) expressed his support of the concept of a uniform world curriculum for secondary-school education. At one point he said:

Is it not possible to see whether we might not develop a uniform curriculum for secondary schools, the same for every country, or group of countries? Such a uniform syllabus should not refer to physics only, but we should have uniform syllabi in every subject. After all, a pupil has only one head, and its receptivity is limited. Physics should be connected to a series of subjects: mathematics, chemistry, biology as well as literature, language, history, and so on. The big pedagogical-psychological problem should be planning a uniform syllabus for all pupils, who are after all, the objects of our teaching.

_____

Chapter 7

Curriculum Innovation in Teacher Education
_____

Although not chosen as a subject for a working group at this Conference, the question of curriculum innovation in teacher education itself was in the forefront of the minds of many participants. In particular, a stimulating paper by G. Marx (Hungary) discussed "An Insoluble Task: Teaching Physics."

Today it is commonplace to speak about the arrival of the scientific revolution. Not only we scientists, but the blue-collar workers, the politicians, even the Masters of Art are forced to take this into account. Many of us, mainly teachers, scientists, and engineers are enthusiastic for the scientific revolution, for we believe in the blessings of technology. We see the realistic promise of a coming golden age. One group of blue-collar workers (the more resilient one) work in the new, booming trades of electronics and automation. Another group (slower and less perceptive) sees a threat in the constantly changing world of modern production. In their eyes the computers are enemies to be destroyed, just as the power looms were for Ned Lud. What about the masters of scientific revolution? Some sacrifice on the altar of the new mystery called science; others see the dawn of a new age of barbarism. They withdraw into impotent opposition. A juvenile and extravagant formulation of this feeling is seen in the slogan "Flower Power." It evokes sympathy when contrasted with nuclear power, but it is retrograde from the social point of view. This slogan tries to drive the human genius back into the pre-human Garden of Eden. The worst attitude of all is the satisfied indifference of those who are ready to accept passively the luxuries offered by technology. They realize happiness through their bellies! In them, technology breeds indolence of the body and the spirit. Indolence is the antithesis of that which has called technology into existence. It is the opposite of creative dissatisfaction.

So technology produces a triple tension:

1.

Its full realization needs increased knowledge. Human knowledge is doubling every ten years, and this has no precedent in history.

Neither the human brain nor the psyche nor the school is prepared
for this information explosion.
2.
The Environment is transformed so fast, the power of man increases so fast, beyond its habitual framework, that moral and
psychological instability appears. We have to speak about aliena-
tion.
3.
Before the over-all victory of the scientific revolution the possibility of a reaction has appeared: killing creative activity by
the passive acceptance of luxury, killing science by one of its
own children.

These contradictions have been produced by the enormous
intensity and acceleration of the progress of science. And the
solution of these contradictions is not to be found anywhere but
through the physics teachers. They teach physics, the fundamental science of mechanical and electrical engineering and the starting point for chemistry and biology. Physics is the best respected,
fastest growing, and, at the same time, the most abstract natural
science. Even modern chemistry and biology may be considered
archaic compared to modern physics. They make use of subject
matter that was discovered by the physicists of forty years ago.
But, into the physics classes crowd the inventions which the scientific revolution has put on the front page of the newspapers,
from the H-bomb to the moon rocket, from the transistor to col-
or TV.

The task before us is to elaborate all these scientifically and
to humanize them psychologically. This gigantic task seems un-
solvable; yet its solution is an absolute demand of society. The
quality of physics teaching is of decisive importance for the future
of our country and for the social happiness of the coming genera-
tions.

If we accept this requirement, the training of physics teachers can be strung on four threads:
1.
Experimental physics. We have to relate a whole series of physical phenomena directly to the world around us, in the hills, on
the streets, and in the factories. We have to face the students
with the phenomena, and we have to put tools into their hands

which work according to the phenomena. Experimental physics
is necessarily encyclopedic and descriptive.
2.
Theoretical physics. The rapid evolution of science ensures that
it is impossible to learn today what one has to teach tomorrow.
It is essential for the teacher to understand the technical products
of the decades ahead of us (the electronic transmissions of odors,
the laser gun) either from books or from obscure hints to be published in the newspapers of 1984. This is possible only if the teachers understand fully the complete system of the laws of nature,
because these future inventions are based upon these laws. Therefore, it is absolutely necessary to teach them theoretical physics
purely and logically. The essence must be taught, the fundamental relations must not be hidden by brightly colored phenomena
or by sophisticated applications. Theoretical physics is by no
means encyclopedic. Theoretical physics is necessarily logical
and puritan.
3.
Special physics. As I have stressed, the teacher must be ready
to understand a fresh science or a fresh technology on the basis
of his fundamental knowledge. This type of activity must be developed by intensive discussion of selected chapters of physics
and technology by means of special courses, seminars, practica,
homework problems, thesis-writing. From all this it follows that
this branch of teaching is definitely incomplete, illustrative, and
readiness-cultivating. The special physics classes are of necessity spontaneously active and facultative.
4.
Teaching physics. The future teacher must be helped to solve that
task in the secondary schools which we have considered to be un-
solvable at the university. He has to teach the whole of physics,
to incorporate it into the culture and the psyche of the individual
and of society. Physics classes are by no means repetitions of
secondary-school textbooks, for these will be out of date when
the present students have become active teachers. The task is
not merely training in pedagogical skills and in the techniques of
experimental demonstration. The physics class is by no means
encyclopedic and complete, because the most complete curriculum and the knowledge of the best professor are unavoidably im-
perfect. Training in physics teaching is a training in pedagogical

skills and social psychology at the same time, developing skill, tact, and a sense of responsibility.

We cannot give up any of the tasks mentioned. Each is equally important. But how are we to incorporate all these into the training of physics teachers? There are different approaches. We show two extreme possibilities:

	Methodological blocks (University)		Thematical blocks (Secondary school)
First year	Experimental physics		Mechanics
Second year	Theoretical physics		Electricity
Third year		or	Heat
Fourth year	Special physics		Quantum physics
Fifth year	Physics teaching		Specialty

The first version (the building up of methodological blocks) is the most common. The Hungarian universities follow essentially this scheme. Within these blocks (experimental physics, theoretical physics, and so on) thematical subgroups are formed (experimental mechanics, experimental electrodynamics, and so forth). In this way the same physical problem (for example, the magnetic field of the electric current) appears twice or even three times during the university course, but is always seen from another point of view. So a risk develops that the physics of the blackboard, of the laboratory, and of the street become isolated from one another.

This danger has made the second version (that is, the formation of thematical blocks) attractive. If in one year the students were kept busy only with mechanics of different types from different points of view, they would be intensively forced into the world of mechanical phenomena. This attractive idea has been widely discussed, but it has not been widely applied. The main objection is that it is very difficult to find teachers able to realize the fourfold program in a well-balanced way. University instructors prefer to be experts in either experimental or theoretical physics; they prefer to be experts either in experimental electronics or in quantum theory but not scientists in the old humanistic meaning of the world. So the university drops the idea of the thematic blocks. But we must not forget that we demand just this sort of balanced completeness from secondary-school teachers.

I am not going to suggest that there are decisive arguments for one or the other structure. One has to leave both possibilities open and also the many transitional forms, by fitting them to the traditions of the university, to the personality of the professor, and to his spirit of enterprise. We must, however, pay attention to the interplay among the different blocks. They must not be in conflict with each other, and none of them must overthrow the delicate balance.

It is not dangerous if one instructor is a structuralist and another is oriented toward a historical approach. It does no harm if one is deductive and the other inductive in approach or if one starts from experiments. But it is unavoidable that the different influences build up a unique world picture in the mind of the student.

In the case of the thematical blocks scheme, this synthesis may be catalyzed by the personality of the professor who teaches the main classes in general mechanics, thermodynamics, and so on. This is much more difficult in the case of the methodical blocks. We have to attain our aim by taking care of individual students. The most effective form of this activity is the formation of small study groups or seminars. The leader of the group helps not only in understanding the main lectures but he guides the students' individual reading and research. This makes the old English system of colleges a reality again.

The job of the schoolteacher is a double one: every day he must build an up-to-date world picture for himself, and he must be able to present this dynamically complete picture to his students. The secondary-school teacher does not live only for the "Experimental Physics" course or for the "Teaching Physics" course of the university but for all the blocks at the same time.

Let me now refer to actual tendencies within these frameworks.

The newspapers write about a spaceship that is navigating on

the far side of the moon, isolated from the earth. This is considered to be the most difficult period of the whole trip. This
dangerous maneuver was unavoidable if one wanted to arrive at
the moon in 1969. "But why is this so difficult? Columbus navi-
gated independently from the home base 400 years ago!" asks
the pupil. The teacher must be ready with the answer. The ocean
is a convenient frame of reference, the sailor can refer both lo-
cation and motion to its surface. But such a reference point does
not exist in deep space. The teacher must explain the idea of inertial navigation and therefore of inertial frames, the theory of
relativity, the relativity of space and time. Is this an unforgive-
able deviation from the subject? I do not think so. The concept of
time has its origin in the fact of motion. This is the view taught
in the theory of relativity, which is a part of university training.
On the other hand the superstructure (the second, the minute, the
hour) is taught in the first grades of elementary school. This is
an absurd situation from the scientific point of view. But the experts are ready with an explanation. The theory of relativity is
too abstract, they say, even for the teachers, let alone the pupils.
    Albert Einstein once asked Piaget, the greatest living author-
ity on child psychology, "How does the small child arrive at the
concept of time?" The result of psychological investigation was
as follows: every child follows the same route from motion and
velocity to the abstraction of time, just as the theory of relativity does. It is the fault of the rigid and formalistic secondary-
school training that makes us consider this natural way of thinking too difficult for the children. Piaget has shown that velocity
is more fundamental psychologically than time is. The secondary-
school training can start with the principle of relativity and follow this with the more sophisticated axioms of Newton. This understanding must be produced in the minds of the pupils. It is a
demand of the age, where newspapers write of astronauts feeling
the state of weightlessness. Incidentally, it is well known that
Robert Karplus teaches relativity (that is, the equivalence of the
different frames of reference) in the fourth grade of elementary
school. Here in Hungary we are experimenting with the teaching
of such fundamental concepts as those of matter, irreversibility,
relativity, and interaction under the supervision of Mrs. Palffy
in Pécs. Such concepts, generally considered to be fundamental,
are usually taught only toward the end of the physics courses.

Let me consider a second example. Every evening the transistor radios are playing in the hands of the young people. We are grateful that they do not turn them on in our classes. Physics teaching has not fulfilled its goal if our pupils do not ask, "What is a transistor? How does it work?" They must be willing to use their knowledge of physics in ordinary life just as they make use of their mathematics when they are shopping. Now the transistor and such other everyday phenomena as the hydrogen bridge in the multiplication of DNA, the tunnel effect in radioactive decay and also in the tunnel diode, the photosynthesis in the chlorophyll of the green plant, all work according to the laws of quantum mechanics. And now we have arrived at one of the central problems of physics teaching. We cannot avoid the conclusion that quantum mechanics should be shifted from the end of the university training toward the middle of it. Chemistry, biology, solid-state physics, quantum electronics, nuclear physics, all depend on quantum mechanics. Quantum mechanics must become a natural way of thinking for the secondary-school teacher of today just as the mechanical style of thinking was natural in the last century. Without training in quantum-mechanical thinking we cannot solve the big unsolved problem of secondary-school teaching—the building up of a useful concept of the atom in the mind of the pupil.

The concept of the atom appears in school courses several times. Elementary chemistry courses refer to rigid atoms. The secondary school-physics course draws on a Bohr-like picture, saying that the spherical hydrogen atom is a flat circle. The main problem is that the classic rigid atom of the kinetic gas theory and of inorganic chemistry is not sufficient for an understanding of the transistor in the physics class. The Bohr model is not enough to understand the formation of the DNA molecule in the biology course. In Hungary this has become the subject of a lively discussion. "What kind of picture should be built up in the minds of people about the atom?" Briefly expressed, this becomes, "How do we bring quantum mechanics into the secondary school?" This is a major pedagogical problem for the coming decades. The necessary condition for its solution is detailed quantum-mechanical training in the education of the teacher.

Let me conclude by asking if it is possible to include a relativistic point of view at the beginning of physics and quantum mechanics in the middle together with a very much freer time for the development of spontaneously acquired learning. Is this not contradictory in the present structure of the secondary school curriculum?

In order to gain time and to develop scientific thinking we are trying to teach science from the first grade in an elementary school. The experiment performed in Pécs under the direction of Mrs. Pálffy seems to be surprisingly successful. In accordance with the results of Piaget about child psychology and the ideas of modern physics, our task is to make the children understand what the concepts objectivity, causality, irreversibility, and relativity mean. I am convinced that science is more primary than mathematics, and we must start and not only finish with science. This curriculum improvement in elementary school may help to ease the burdens of the secondary schools.

Many weak and mediocre students have already left Budapest University. The truly talented ones are always in the minority. But even the weak students have become surprisingly useful and conscientious teachers in school. Let us trust them. They are able to learn and to teach the ordinary material of elementary physics courses by themselves. In mathematics courses in the University we do not teach the multiplication table. Why then do we teach geometrical optics, statics, and other examples of doorbell physics there? We may forget about these things. Rather let us teach them about those subjects that will appear in the newspapers and in the schoolbooks of 1984 when they will be teaching in the isolation of the country schools. Let us teach them science as a dynamic, ever-developing unity. Let us teach them the schoolbooks of tomorrow.

Whereas G. Marx was largely concerned with a reshaping of the physics course content for future physics teachers, J. B. Cross (U.S.A.), presenting a plan for " A New Four-Year University Curriculum for Future Teachers of Physics and Chemistry in Secondary Schools," proposed a complete restructuring of both the science and the education content of the course:

One of the most critical problems in science education today is the shortage of qualified physics teachers in secondary schools. This problem is further aggravated by the fact that the develop-

ment of new curricula in secondary-school science has not been
followed by corresponding changes in the university curriculum
by which science teachers are trained.

The new science curricula, at least in the United States differ from the old ones not only in content but also in methodology
and approach. Most teachers teach the way they themselves were
taught rather than the way they are told to teach. Thus it is not
enough in creating a new program for physics and chemistry teachers just to change the content of some of the courses to be taken
at the university. The entire mode of instruction has to be adjusted to the new attitudes and skills the future teacher should
acquire.

With these two goals in mind, the Physical Science Group,
successor to the Physical Science Study Committee of Education
Development Center (EDC), is in the process of creating new curricula in physics, mathematics, and chemistry for the training of
science teachers.

Teachers teach the way they were taught. Therefore, the
mode of instruction in the program must be consistent with modern ideals of good school-science teaching from the first day on.
This will manifest itself, among other things, by (1) the large
fraction of time devoted to laboratory work, (2) the use of the
laboratory as the major tool to the understanding of basic concepts, (3) the preference of discussion groups over lectures,
which makes the student active also in the classroom, and (4)
the encouragement of continued study after graduation. Students
will be called upon to engage in individual study from books and
resource materials. Items (2) and (3) deserve special attention.

The laboratory work of the students will be handled in a way
that is uncommon in nearly all universities. Most of the experiments in the program we plan will not be confirmation experiments of some general law already known to the student nor the
measurement of some physical and chemical constants already
known. The student will go into the laboratory without knowing
what the results of his experiment will be. After completing his
work his results will be pooled with those of his classmates in
the form of graphs, histograms, or tables that strengthen the
validity of the results of the experiment. This saves a great deal
of time, since a single student would have to make many measurements to arrive at the same degree of validity as that obtained
from the combined results of all the students.

In the standard undergraduate courses for chemistry and physics majors, teaching by lecture is the rule. We believe that a heavy emphasis on lectures is unwise at the secondary-school level. It is only in class discussion, with the teacher as discussion leader and guide and the students actively involved, that, we believe, the most effective learning takes place at the secondary school level. It is only by conducting undergraduate courses along these lines that future teachers will learn how to teach this way efficiently. Furthermore, the practice the students will get in expressing themselves clearly and in understanding their classmates' explanations and questions will help them conduct meaningful classes once they become teachers. Although this may seem to be an inefficient mode of teaching science at the university level it must be remembered that a future schoolteacher does not need to cover as much content in science courses as a student seeking a degree in physics or chemistry. In addition, combining chemistry and physics in a unified curriculum will save much time by avoiding redundancy in the two sciences.

Student practice teaching in secondary schools will start in the second year. The largest possible number of third- and fourth-year students will also be called upon to assist in the teaching process in both the laboratory and discussion sections of the first-year courses.

While the science and mathematics subjects are planned in a definite sequence, the other topics listed here can be taken at a variety of times:

A.

Physical Science I: About 60 percent laboratory work, 30 percent discussion and problem solving, and 10 percent lecture.

As these figures indicate, we wish to depart at the beginning from the traditional lecture-centered college course and give the students their first learning experience at the college in the same way we expect them to work with their own students in the future. There will be a large number of contact hours per week in this first course. This has a twofold purpose: it will rapidly diminish the importance of the different backgrounds in science with which the students are entering the program, and it will help the students to move into more advanced topics sooner.

The theme of this course is the study of the basic properties of matter leading to the establishment of the elements of the atom-

ic model. It will follow the college edition of the physical science
textbook written by the EDC Physical Science Group.[1]
B.
Physical Science II: About 60 percent laboratory work, 30 percent discussion and problems, and 10 percent lecture.

This course is a continuation of Physical Science I and deals
primarily with the connection between electric charge and atoms,
leading into a study of the various forms of energy and the conservation of energy.
C.
Experimental Physics: 50 percent laboratory work, 30 percent
discussion, and 20 percent lecture.

This course will be based on the college edition of the Physical Science Study Committee (PSSC) textbook, College Physics.[2]
The basic theme in this course is the kinematics of waves and
the dynamics of particles including charged ones, leading to the
quantum-mechanical picture of the world. The emphasis is primarily on fundamental laws applied to single particles.
D.
Experimental Chemistry: About 50 percent laboratory work.

The first topic will involve the direct analysis or preparation in the laboratory of some compounds. Once the ideas of reactions and products are introduced, the question of how the
course of reactions can be manipulated will lead to an empirical
study of equilibrum and Le Châtelier's principle.

Further investigation into temperature effects on equilibrium will lead to a correlation between the heat of reaction and
equilibrium constants. This will combine with the development
of entropy in the PSSC physics course to provide a quantitative
and nonabstract basis of thermodynamics.

----

1. College Introductory Physical Sciene. Prentice-Hall, Inc.,
New York, 1969.
2. Physical Science Study Committee. College Physics. Raytheon
Education Co., D. C. Heath, Lexington, Mass., 1968. This text
and the accompanying laboratory manual has the same general content as the second edition of the PSSC school text except that much
of the content of Part I of the school text has been deleted and new
topics were added. The new topics include the material on angular
momentum, statistical mechanics, relativity, atomic and nuclear
physics contained in the PSSC "Advanced Topics" supplement.

Further ideas of control of reactions will be pursued in the study of electrochemical cells, which involves oxidation and reduction mechanisms as well as thermodynamics. Energy storage and retrieval by chemical means will be discussed in terms of biochemical examples as well as in terms of electrochemical cells.

Macroscopic evidence about the structure of molecules and bonding will be pursued historically, following the reasoning of Couper and Kekulé. These models will serve as an introduction to the study of reaction kinetics for clues about microscopic reaction mechanisms. Recent experiments on reactions in molecular beams will be used as examples.

Almost all theoretical ideas will be presented as models proposed to correlate student experimental findings, following the pattern established in the earlier courses. Basic laboratory skills will be acquired as needed.

E.

Chemophysics: About 30 percent laboratory work.

This course, coming after the courses in experimental physics and experimental chemistry, has the purpose of tying together macroscopic and atomic properties. On the static side this will include a correlation of cooperative properties such as electric and magnetic constants, index of refraction, and heat of sublimation, with shapes of molecules, energy levels, and interatomic potentials. On the dynamic side the course will take up the study of chemical and nuclear reactions from both the thermodynamical and statistical points of view.

It should be pointed out that the laboratory experience the student will acquire in these science courses will be at least that of a regular Bachelor of Science in physics or chemistry.

F.

Theoretical Physics: Lecture and problem sessions.

While a high-school teacher need not be an expert in lengthy calculations, he must be conversant with the basic language of theoretical physics. Therefore, the emphasis will be on understanding the physical content of basic mathematical terms such as gradient, divergence, and Laplacian. This will lead to a unified view of physical phenomena that otherwise might appear unrelated.

G.

Mathematics.

The mathematical skills a science teacher must possess and be able to transmit to his students can be broadly divided into two categories: what to do and how to do it. The first category includes the translations from daily language; the second includes the actual techniques of solving well-defined mathematical problems. In the mathematics courses we begin with such basic concepts as difference, relative difference, and ratio as modes of comparison. Emphasis will be placed on methods of estimating and approximating. The material will cover all the mathematical tools needed by a physicist or chemist at the undergraduate level, but the emphasis will be on applications rather than on formalism. Much of the content will be geared to the specific mathematical needs of the students in their science courses and, in the first year, to the problems encountered by teachers in schools in using mathematics in the teaching of physics and chemistry.

H.

Workshop Practice.

All science teachers should be able to take care of minor repairs for their laboratory apparatus. Teachers with original ideas should be encouraged to build their own apparatus. Therefore, we think that a sound course in shop procedures is an essential part of the education of a science teacher. This will include basic wood, metal, and glass work and also familiarity with electrical and electronic measuring devices and elements of vacuum techniques. Many of the exercises will be taken out of actual situations, such as diagnosing common troubles and fixing them.

I.

Testing and Measurement.

Courses aimed at the understanding of principles and their application to new situations often fail because the achievement tests that go with them are geared to rote memory. Teachers must therefore be equipped to develop tests on their own which will be consistent with the goals of their teaching.

In this course the students will have an opportunity to develop tests, including the design of multiple-choice tests and their evaluation. They will also practice the design of laboratory tests.

In addition to the experience in developing tests derived from this course, students will begin, as early as the first year, to write tests covering material recently studied. These tests will be taken by fellow students and their effectiveness discussed in class sessions.

J.
Language.
The purpose of the English course is to enhance the ability
of the teachers to express themselves clearly both orally and in
writing. The work in English will be centered around their science activities and will include the improvement of laboratory
reports, book reports, and lectures by students on scientific matters from the point of view of language. Our observations of teachers' writing and speaking in the classroom convinces us that this
is a very necessary aspect in undergraduate education of future
teachers.
K.
Practice Teaching.
Practice teaching in schools can start early and be spread
over several semesters. In the last year it can include also assisting the teachers of the first-year courses of this program at
the college itself, thus helping to hold down the ratio of instructors to students.
Throughout the four years there will be close cooperation
with the faculty of education. Members of the faculty will observe
students in action during discussion meetings in the various
courses listed and from these observations be able to help the
students directly in the art of teaching.
L.
Educational Psychology.
M.
Electives.
We hope that under electives some students will choose a
foreign language, both to broaden their cultural horizons and to
enable them to gain teaching experience abroad at some time in
their career.

There are a number of countries in which the prospective
physics teacher follows the normal university course in physics
and then takes a further year of study in the field of education before entering the secondary school. Typical of these is the United
Kingdom. In his paper "Towards a Satisfactory Model of the Education of Physics Teachers," B. R. Chapman (U.K.) took a point
of view that contrasted sharply with those expressed by G. Marx
and J. B. Cross:

I believe we must see the education of teachers much more
as a sociopsychological experience than as an academic one. In
all teaching we despair of how little our students take in or benefit from our bestowed wisdom. How much more true this must be
of educating for teaching than of almost any other profession.
From an experience of sixteen or more years of being a pupil, in
classes of between 20 and 100 fellow pupils, one is catapulted in
the short space of one year into the role of the teacher responsible for the learning that goes on in these groups.

The result of this change of sociological role, from that of
student to that of teacher, can be traumatic. Unless one's personality can come to terms with this change, the likelihood that
a series of tips for teaching, or a systematic attempt to cover
identifiable skills that a physics teacher needs, will be successful is rather remote. Conversely, if one's personality can as-
similate this role change it is likely that the actual course content is not really of paramount importance in making the course
a success for that teacher.

It is this sociological element in the education of graduate
physics teachers which is to me the most significant. It does impose realistic limitations on what we can hope to achieve, and,
perhaps most important, it does suggest the kind of emphases
that are most likely to be productive in the education of graduate
physics teachers.

If this analysis of the task in front of us is acceptable, then
the development of a satisfactory course becomes much more a
matter of developing an appropriate learning environment than of
content. Indeed, an extreme view might argue for a totally un-
structured course in which no topics were predetermined. Such
a course might develop if the tutor concerned determines that the
optimum learning environment is some form of small group structure out of which all decisions about course content and activities
will arise. The tutor may totally reject the role of supplier of
predigested information and advice. Instead, he will rely on the
outcome of group discussions and experiences to produce the desired outcome of the course, whatever that desired outcome might
be.

Unfortunately there are very few of us engaged in physics-
teacher education who are strong willed enough to do this. We all
secretly believe we know how to do the job and that all we have to

do is to find an effective means of transferring this know-how to our students. Yet often what we do is to produce the same kind of total lack of communication as is present when the erudite low-temperature physicist lectures to the school science society.

The real difficulty with this group approach is that graduates coming into teaching are totally ignorant of what methodology is all about. By and large they have never looked at a specific heat experiment from any other than the students' point of view. "What does it look like to the teacher?" is a question that has never occurred to them. It is perhaps much more important to get them to ask this kind of question for themselves than to ensure that by the end of the course they have satisfactory answers to it.

It would seem therefore that a minimal amount of course structuring is necessary. On the whole this should be in terms of letting the students know the possibilities within the course rather than a structured approach to what to do about them. For example, the use of a video recorder and camera for microteaching is desirable; the structuring of this method is not. Small group teaching in school is desirable; the structuring of this experience in terms of stating objectives, methods of teaching, evaluation, and so forth, is not, unless the groups of students decide that such structuring was of value to them in their own teaching.

Given this, the course, although superficial about methodology, becomes in reality a reorientation exercise. It is the methodologist's first task to provide the environment for this reorientation from student to teacher to take place. It is his second task to monitor it in such a way that he can be of most use to individual students who find this transition difficult. He does not worry overmuch about the state of a student's knowledge about laboratory design, evaluation, curriculum theory, educational technology and so on, although he will probably be disappointed if the student shows no interest in anything!

This must be the essence of any course of initial education and we must not lose sight of this in a plethora of advice about course content. Our expectations must not be too high because the personalities with which we are concerned have had twenty years of previous experience built into them. How much we can do in one year to help or hinder the intending teacher when he first faces the tensions of the classroom is open to doubt. Whatever we do achieve will be because we have through our courses given him some confidence and some self-awareness in facing up to his task.

A new tool for use in teacher-education programs at both the initial and in service levels was described by E. M. Rogers (U.K.) in a paper on "The Use of New Examinations and of Examination-Construction Seminars in Curriculum Revision and in the Training of Teachers." In effect, Rogers allows an examination-question-making seminar ("shredder") to develop into a much wider discussion. He says that it was a surprising discovery that

we can generate a valuable discussion of teaching philosophy by starting with an examination-making seminar. As I practiced it there and in later seminars, I found my work as chairman enabled me to steer the discussion in a manner that makes it a very fine way of getting teachers to understand the aims and methods of a new program. As chairman, I have at first to encourage the discussion and criticism of questions; but then I retire into a less active position, merely adding comments on aims and on feasibility of suggestions. Teachers invited to a meeting to discuss their philosophy of teaching usually hesitate or else launch into unrealistic aims. But teachers invited to an examination-making seminar soon discuss realistic aims.

When you are making examination questions, you have your eyes focused on a very practical matter. It appears as though this would be a very specialized way of training teachers. But notice its chief characteristic: it is training by doing. It is the teachers round the table who make the questions, talking to each other, discussing their views, and developing their ideas. I have every hope you will try it yourselves with a group of colleagues, the more varied in their views the better. Ask them to make examination questions for you; bring them together the next day, and then hold such an examination-question-making seminar as I have described. My hope is that teachers, by talking among themselves about what they teach, how they teach it, and what their aims are, may educate the children in a new way, a way that will ensure that the parents of the next generation will say to their own children, "Go and do physics and enjoy it; physics is interesting experimentation and clever thinking put together; go and enjoy it."

Answering a question about the advantages of the oral examination, E. M. Rogers said:

In some ways I regard oral examinations as of great value. But they are extremely low in reliability. I sit in oral examinations with two other examiners, and the three of us never seem to agree except when we get a remarkably good scientist or a remarkably bad one. The point about such examinations is that they are a conversation with the student; therefore we can supply him with help and that help makes him able to tell us more about him. But this element can be brought into written examinations. If you write a printed examination with much more information in the questions, you make it much harder for the candidate to escape with too vague a comment. You can practice the game of throwing the ball to and fro between examiner and examinee even in written questions to some extent. Of course, if you work in a teaching system where continuous assessment is possible, much of the examining will be of the oral type and done by the teacher himself. In such cases, the needs of the university examination can be met by a calibrating examination of some kind. This is the practice in Sweden, where the schools are calibrated rather than the individual students.

Experiments are being carried out at the University of Surrey, England, which involve the students in self-teaching situations. These include the issue of prepared notebooks for the lecture courses, self-tests, tape and tape/slide presentations of lectures, and a self-service laboratory. Of this latter aspect L. R. B. Elton, P. J. Hills, and S. O'Connell (U.K.) in a paper called "Self-Teaching Situations in a University Physics Course for Secondary Physics Teachers" have this to say:

Traditional laboratory work occupies a considerable part of the students' time, normally in periods of 3 to 6 hours. It is usually justified in terms of a number of objectives, such as relating theory to practice, developing practical skills, writing clear and concise accounts, encouraging creative thought, aiding the understanding of concepts, and so forth. It is, however, rarely clear which of these are meant to be achieved by any one particular item of practical work, and the very general dissatisfaction felt by staff

and students everywhere with orthodox laboratory work is almost
certainly due to this confusion of aims. We therefore set out to
separate these objectives by devising types of experiments that
will achieve a small number of them while consciously excluding
the rest.

The objectives considered have been derived from those suggested by answers to questionnaires given to students and staff
and are listed in Table 1. They have been divided into two broad

Table 1
Objectives and Laboratory Experiments

Objective	Type of experiment
(a.) To acquire practical skills (b.) To develop measuring techniques (c.) To gain knowledge of facts (d.) To aid understanding of principles	Short, **single-pur-pose experiments with programmed script**
(e.) To encourage creative thought (f.) To develop judgment in experimentation (g.) To develop self-organization (h.) To write clear and concise accounts (i.) To learn how to acquire knowledge and skills for an immediate purpose (j.) To develop an appreciation of the limi-tations of oneself and of available re-sources	Long, open-ended experiments

main groups, the cognitive and affective domain, respectively, and associated with two types of experimental work, referred to as the single-purpose and the open-ended experiment, respectively. The former typically lasts about half an hour and is a structured experiment with a programmed script, which makes it possible to lead the student by a series of small steps to the desired objectives. This approach is suitable when objectives can be precisely formulated, as is the case with (a) through (d) in the table. Objectives (e) through (j) are much less well defined and relate more to real situations, as met by professional physicists in their work. For that reason they are most readily met by putting the student into a near-real situation, exemplified by an open-ended experiment. This is an investigation that, apart from a minimum of initial instructions, is entirely student directed. Within reason, no time limit should be set for its completion.

We have concentrated on the development of single-concept experiments, designed to achieve the required objectives with the use of a minimum of associated knowledge and skills. For instance the setting up of apparatus is reduced, where appropriate, to a few initial adjustments, and calculations are included only when necessary to reinforce an illustration of relationships or to provide the satisfaction of obtaining a numerical answer. Whenever the experiment is intended to reinforce lecture material, it takes the form of a demonstration carried out by the student himself, so that he acts as his own demonstrator. This self-teaching aspect is common to all such experiments, which are complete in themselves and do not require the presence of the teacher.

It should be noted that a laboratory with self-service components is not a means of eliminating demonstrator-student contact. It does, however, eliminate many of the mechanical and repetitive tasks that at present have to be performed by demonstrators. It could leave them free to concentrate on teaching those aspects of experimental physics that require the free interchange of ideas between student and teacher; this can take place naturally in the laboratory situation.

Our investigation is proceeding in three parts.
1.
Single-concept experiments linked to the syllabus. So far one experiment, the operation of the cathode-ray oscilloscope, has

been designed in connection with objectives (a) and (b), while six others relate to objectives (c) and (d). These are

Potential well and barrier
The gyroscope
Measurement of e/m for electrons
Waves and interference
Collision in two dimensions
Equipartition of energy.

Only the first four of these employ actual apparatus, while the others are simulated experiments or use existing film loops. Students have stated that they like the style of experiment because the objectives were clearly stated, the apparatus always worked, there was no writing up afterward, time was used efficiently, and there was more time to think. Staff involved have confirmed that students seemed to profit more in acquisition of knowledge, skill, and understanding than from conventional experiments.

    In connection with the use of film loops and slides in simulated experiments in elementary dynamics, a teaching booth has been designed that incorporates back projection on to tracing paper, so that a student can record the progress of bodies in motion, using a clock and the stop-frame control of the loop projector.

2.
Different means of presenting a single-concept experiment. The question of which method of presentation is most suitable for a single-concept experiment was investigated for an experiment on the photoelectric effect. Four methods were used:

A normal, conventional, practical script used in conjunction with the bench apparatus.

A programmed form of the practical script used with the bench apparatus.

An audio program, which used a tape recording and duplicated material with the bench apparatus.

A simulation of the experiment, using a combination of 8 mm sound film and a tape/slide presentation.

The objective in each case was to achieve an understanding of the effect and not to acquire experimental skill. Tests were given to establish the students' state of knowledge before and after the experiment and to assess their attitude toward the various modes of presentation.

Conclusion

The purpose of the present investigation is to provide a series of prototypes for components of a teaching and learning system. **In** a number of cases we have found that different students prefer different methods to reach a given objective; this may well indicate that these should indeed be provided. On the other hand it is not obvious that students will know in advance which method they prefer in any particular instance.

Chapter 8

The Technology of Physics Education

The working group on The Technology of Physics Education[1] felt
that it would be more useful to survey this subject rather than to
present to the Conference as a whole any specific guidelines.
Their report is as follows:

Developments of modern technology, of new media of communication and of a multiplicity of aids to teaching and learning
have and will increasingly exert a considerable influence on educational practice. Starting with a variety of devices such as films
and light projectors, it has led to the development of teaching and
learning units. This in turn has led to the idea of systematic approaches to the teaching and learning process in which the formulation of objectives and the provision for appropriate evaluation
procedures and feedback play a crucial part in the construction
of an educational system. All this is usually referred to under the
title "Educational Technology," although the term is coming more
and more to be restricted to the full systems approach to education.[2]

The reason for this is that it has become apparent that the
straightforward application of technological hardware has proved
of variable value, and that the creation of the associated software
raises the kind of problem that may best be solved by a systematic
approach. The "packages" that have resulted from this treatment,
which, it must be stressed, is applicable to only certain parts of
the educational process, contain hardware as appropriate, but the

----

1. Members of the working group which discussed this subject
were B. R. Chapman (U.K.), Chairman, L. R. B. Elton (U.K.),
Rapporteur, Mrs. M. Bochenek (Poland), R. David (France), W.
Eppenstein (U.S.A.), K. Hinst (O.E.C.D.), W. V. Johnson (U.S.A.),
J. A. Kobus (Netherlands), M. Lemaître (Belgium), V. Metev
(Bulgaria), D. Pescetti (Italy), O. P. Puri (U.S.A.), Mrs. E. A.
Shekuteva (U.S.S.R.), R. S. Tilton (U.S.A.), H. Volz (Federal
Republic of Germany).
2. A glossary of technical terms used in this chapter is to be found
at the end of this chapter.

latter no longer forms the essential part of the innovation. Packages may be constructed to help the teacher and to free him from
certain tasks in order to give him time for others, and they may
cover a whole course or only a small part of one. They are particularly suited to encourage active participation by the student
and to allow him to be treated as an individual. Packages are thus
in line with current trends in education. However, it should always be borne in mind that the same methods when used incor-
rectly, such as the mass treatment of passive students, lead to
unfortunate results.

Potentially, one of the most versatile pieces of educational
hardware is the computer. Although at present its use may in
many instances be prohibitively expensive for schools, teachers
trained in its use today will be aware of its potentialities for the
future. However, even today schools are beginning to be equipped
with on-line terminals, including in some instances audiovisual
displays. These are ways in which interaction between the student and the computer becomes an exciting reality. One use of
the computer in physics is to extend the mathematical powers of
students at every level, and here it has been found to be a most
valuable tool. In management, including the processing of examinations, it is also beginning to play a role in school systems
all over the world. There are many other possibilities on which
current work is being done and which will have relevance for physics education.

The initial success of much of what goes under the title of
Educational Technology, including the computer, has at times
led to improper or excessive use. Educational Technology must
always be seen in its pedagogical context and be linked to other
methods of teaching. At all times it should be the servant, not
the master.

The Use of Educational Technology in Teacher Training

In the pedagogical part of teacher training, Educational Technology plays both a general and a specific part. The use of closed-
circuit television, video and audio tape recording, and more recently of the computer for data retrieval and the simulation of
classroom situations are equally applicable to all subject disciplines. The use of the mass media of communication and of cor-

respondence courses to supplement more usual methods of teacher training must also be considered, provided they are linked
with teacher group discussions. For other, more specific uses,
the cooperation between the subject specialists and educational
technologists is required. The modern education of physics teachers is leading to the creation of packages of a modular nature
which can be fitted into training courses, but as yet there do not
appear to be any fully developed systems. Attention is also drawn
to the particular value of video recording for making a student
aware of his own teaching performance as well as of that of others.

Awareness of the Potentialities of Educational Technology

An important aspect of teacher education should be to make students aware of the potentialities of Educational Technology in the
classroom. This should be provided in the technical part of pre-
service training. In this way students will learn not only its value
but also its limitations, such as that simulated experiments on
film loops are rarely as effective as real experiments done in the
laboratory. In the pedagogical part of a student's education it can
be done through the teaching of techniques in the use of hardware,
the creation of software, and perhaps most importantly through
the teaching of methods of assessment of students and the evaluation of innovations. This should be attempted even if it can only
be done rather superfically in the time available.

It is also desirable that in-service training should have a
component of Educational Technology in its curriculum, partly
as a separate course, and partly interwoven into other aspects
of the in-service training, such as with the presentation of new
subject matter or new methodologies.

For both pre-service and in-service teacher education, the
provision for communication centers associated with each teacher-
education institution is vital. As the book is the oldest, and in
many ways still the most important tool of Educational Technology, these centers should in general grow out of the libraries of
the institutions and form an extension of them.

Of special importance to the teacher in the field of Educational Technology is the provision for the training of technicians
and other ancillaries. At present it is rare for them to exist in
the schools of many countries. Although the training of techni-

cians falls outside the scope of teacher education, it is very important to the general acceptance of most of the techniques of Educational Technology.

Special Problems of Developing Countries

In developing countries the use of Educational Technology may well be of crucial importance, and it is particularly necessary here that the investment of scarce resources in educational hardware be undertaken wisely and with due deliberation. In this connection aids need to be developed which are technically simple to use; for example the cassette type of tape recorder is much more widely applicable than the reel type.

The use of self-instructional courses and of educational television is of value both for pre- and in-service education of teachers, and particular attention should be paid to the need for training teachers who can use the mass media and other tools of Educational Technology. There is also a great need for resource centers that can serve large areas.

Conclusion

The working group did not attempt to survey the whole field of Educational Technology. However, even the most cursory investigation showed that the pace of development differs substantially between different institutions and countries and that a great effort will be required to spread even the existing developments of Educational Technology that have been found successful in some places to those that are not using them at present. However, this effort must be made, for the teachers we train today will be in the schools for forty years to come. Long before then, Educational Technology, which at present is still in a very early stage of its development, is likely to play a leading role in the educational process, and teachers who are ignorant of it will have failed to master their profession.

The paper submitted to the Conference by K. Hinst (O. E. C. D.), dealing with the whole area of the technology of physics education, was particularly instructive. It was entitled "Educational Technology, Implications for the Educational System, National Poli-

cies, and the Role of the Teacher in an Individualized Learning
System" and is reproduced in part as follows:

Educational Technology: A Definition

The past twenty years have witnessed in Europe as much as in
the United States the development and use of new ways of communication by means of technological devices that have not only
exerted a tremendous influence on our behavior in the professional world and during leisure time but have also had an impact
on ways of thinking about education. They have revolutionized the
possibilities of conveying information and knowledge to students
in a way that can be compared only to the invention of printing
five centuries ago.

    Already, audiovisual aids, such as loops, tapes, and so on,
have influenced standard educational practice, and the possibility
of change in teaching techniques is even further advanced by the
introduction of such media as television and, to some extent, computers.

    For some time now it has been common practice to apply the
term "Educational Technology" to those activities that are concerned with the introduction and operation of the various kinds of
"hardware," that is, the various media ranging from audiovisual
aids to the very complicated new media systems. This concep-
tion of the term has been propagated by those who are concerned
with such machines mainly from a technological point of view and
who are thus less concerned with the educational implications of
their utilization. It is this perhaps that has been the major factor
to date in the failure of the new media to fulfill the high expectations that were linked with their introduction.

    This concept of Educational Technology (ET) proved to be
too narrow, and the most recent thinking about ET divorces it
from hardware and conceives it as a systematic approach to the
teaching-learning process. To understand this change to a wider
concept of ET, one has to consider the main factors which account for its emergence:
1.
Given the availability of hardware, the concern of the education-
alist in practice is now focusing on the production of adequate
software, that is, the problems of content and presentation of the

message, which are at present lagging very much behind, but which form the core of the educational problems if media are to be of any use at all.
2.
The problems of mass education, the increasing numbers of students, the prevention of dropouts, the dissatisfaction with the present efficiency of learning, and the cry for individualization and democratization of the teaching-learning process have all led educationalists to question the adequacy of the traditional learning system. The result is an increasing preoccupation with, and a fundamental reconsideration of, the total teaching-learning process in order to see which new ways would better correspond with the individual social and economic needs of education in our time. It is against this background of pedagogic and didactic thinking that questions about media utilization are seen nowadays.
3.
This widening of the concept coincides with a general trend toward a more scientific and methodological approach to educational problems, a trend which could have been observed in other social fields and which is now taking place in education.

The advent of a systematic approach to the teaching-learning process assigns to ET the development, production, and evaluation of learning systems. This materializes in the effort to produce prefabricated course materials, that is, objectified learning units that are independent of the constraints of time, space, and personnel with respect to their actual utilization by the student. Such learning systems make use of a combination of various media that include print, in other words multimedia systems or learning packages that might comprise television, radio, learning laboratories, and programmed correspondence material or other combinations. In other words, the single medium is no longer regarded in isolation as a panacea but, as we become aware of their particular advantages and disadvantages, the question becomes one of deciding which medium can best be used for achieving specific learning goals and how can they be optimally combined in a learning system. Second, objectified learning systems call for new teaching techniques, such as continuous and integrated feedback, learning by doing, stimulating motivation on inquiry behavior, self-pacing, and so on. The emphasis of both

is toward individualizing learning, so that it corresponds better
to the different learning habits of the individual students and is
also more independent of the constraints of conventional classroom teaching. The contribution of the new package approach to
the improvement of teaching and learning can be seen primarily
in individualizing instruction, better coping with mass student
enrollments, and democratizing the learning situation.

Implications for the Educational System

The approach to a systematic reconsideration of the teaching-
learning process and the design of learning systems so far have
revealed that one cannot deal with these questions without at the
same time considering the conditions that set the framework of
the teaching-learning process. Factors such as the present skills
of teachers, the nature of the accommodation available, the rigid
timetables for schools and classes, the vagueness of present curricula in relation to educational objectives, all these influence
the process itself. Unless they are readjusted to the new ways
of instruction, it is unlikely that innovations will succeed. It is
here that a systematic approach will provide the basis for more
strategic thinking among educationalists, be it at the classroom
level, the school or community level, or at the regional and national levels of our educational systems, by focusing more closely
on the frame factors of the teaching-learning process while looking at this process at the same time.

The dissemination of the ET approach would imply that both
researchers and teachers will have to change their understanding
of their respective fields; that is, research and teaching will have
to move closer together and the implications of their interrela-
tionship considered much more in each field. Too often research
is conceived from an ivory-tower position by the researchers
themselves, and little concern is shown for how it could and should
help educational practice. However, unless one withdraws to an
individualistic, and primarily contemplative, philosophical attitude which would claim little impact on reality, it is difficult to
deny the inherently "applied" nature of any social science. On the
other hand, even empirical researchers often tend to lose sight
of the actual problems and pursue their research merely for the
sake of research. The consequences of this insight have not yet

been drawn. It is one of the potentials of educational technology
to make educational research more practice oriented, that is, to
test and validate theoretical considerations and research by applying it to the solution of the day-to-day problems of the teaching-
learning process. For teachers, on the other hand, the acceptance of educational technology will introduce a more sophisticated
way of carrying out their functions, a greater effectiveness of
their endeavors, and a closer cooperation between learning, teaching, and research work. In other words, educational technology
is introducing the concept of Research, Development, and Application as an essentially integral tripartite element in any educational system. It is time that this potential impact is fully realized and firmly implemented in our educational thinking and in
our educational practice.

What is needed at this stage is a comprehensive survey of
the problems from an ET point of view, indications of how these
problems can be solved, and an elaboration of alternative strategies for the implementation of ET at various levels of the educational system. Such an exercise would make it possible to provide educational authorities with the necessary information for
their decisions and long-range policy considerations. To date,
very little has been done along these lines. Reforms are still introduced in a piecemeal fashion which, as has been shown in the
past, leads more often to failure than to success.

Implications for National Policies

The work of the Centre for Educational Research and Innovation
at O.E.C.D. has revealed that many of its European member
countries are increasingly concerned with the question of insti-
tutionalizing ET into their educational systems. However, because of the incoherent and inconsistent information about its nature and the implications of its implementation, governments find
it difficult to adopt any long-range policy.

To understand the policy implications of ET, one has to see
its impact in the general context of reforms and innovations that
are at present taking place in education. The preoccupation with
the teaching-learning process constitutes one of the elements in
combined efforts to update education in order to meet the needs
of modern society. As ET is aimed at the microsystem in educa-

tion, namely the interaction process between teacher and learner, upon which all other systems are dependent, it forms the core at which all our educational planning, educational management, or curriculum development will finally have to prove its validity by the effectiveness of the learning outcome in meeting individual as well as societal needs. To achieve this, it would seem unwise to expect any lasting results from innovations in the surrounding fields if one does not at the same time introduce the necessary changes at the base of the educational system itself.

Moreover, ET must be seen in a line of efforts which shift our innovative activities from the mere concern with the quantitative and structural changes in education to an emphasis focusing at the same time on the qualitative, procedural, and content aspects. This change of focus (the necessity of which has been widely publicized by recent student unrest) is only beginning now, but there can be no doubt that this trend is spreading rather rapidly. It is not only educationalists who are becoming increasingly involved in these problems but also national governments, which will have to take the necessary decisions to warrant the contribution of these efforts to educational practice.

National authorities will have to make decisions on how to institutionalize ET. It contributes to all levels of the formal educational system and opens up new possibilities in "further," "adult," and "recurrent" education, which are growing in importance. However, its implementation at these different levels will call for different strategies. Effective progress in this direction will depend on whether concomitant changes in the "frame" factors of the teaching system can be introduced.

Thus, to establish the right mechanisms to improve and update the teaching-learning process, governments would have to invest additional human, material, and financial resources.

However, in order to set up any long-range policy, educational authorities need to be provided with solid information about the complexity of the problem and the consequences of the various actions to be taken. This task cannot be carried out only by educationalists or representatives of the decision-making level alone, but calls for the combined efforts of both.

Implications for the Role of the Teacher in an Individualized Learning System

Any strategy for implementing the production, utilization, and evaluation of new learning systems must accept that the element "teacher" is the crucial point for developing the strategy. The introduction of modern learning techniques and styles will affect the position of the teacher in the classroom in a way that will call for a reconsideration of his role and his various functions. The common denominator under which these changes can be regarded is the change in focus away from teacher-based systems. At a time when teachers are becoming more and more self-conscious about their social status, such a change in focus is likely to create resentful attitudes, especially if teachers do not adequately understand the implications of the change. The fact that even today the majority of teachers still adopt anything from a reserved to a hostile attitude toward any media innovations is an unfortunate example of what can happen. To surmount this resistance, this change of the teacher's role and functions in the classroom must be viewed against the background of his social role and present status within the society. Any strategy that does not take this complex relationship into consideration is jeopardizing its results.

Some considerations of the teacher's role made from the point of view of changing the classroom to a learner-based one, are now suggested. For convenience, the approach of frame factors in relation to the teacher, as outlined in the O.E.C.D. report "Educational Technology — A Systematic Approach to the Teaching-Learning Process," will be used. According to it, the teacher's role and the frame factors he represents at the classroom, school and community, regional and national levels, will be examined.

Frame Factors Related to the Teacher Inherent in the Basic Teaching-Learning Situation

Knowledge and skills What does the individualization of learning and the production of prefabricated course material really mean

for the teacher? Certainly not that such packages can take over
the whole learning process. Even in the future, learning will be
achieved by a variety of methods that assign a central position
to the teacher—a position, however, in which at least one function is considerably changed: the conveyance of information.
Teachers today are still carrying out predominantly this one function, though many people (and there is evidence to support this)
believe that they are least suited for it. It is precisely in this activity that media—and we include under this term anything from
printed material to complicated machines—can take on a major
part of the burden, leaving the teacher free to perform a truly
pedagogic role. The latter has so far been a useful reference
point on which to hang professional ethics, rather than a description of reality. In an individualized and democratized teaching
process, the teacher's job will be to promote the generalization
and transfer of knowledge, as opposed to the mere learning of
facts, and to foster creative activity. He will continuously observe the progress of each student and intervene discriminately
wherever there are obstacles. His pedagogic responsibilities can
thus be aimed more at the development of the student's individual
personality and behavior. In taking on the role of a counselor, he
must necessarily work in close cooperation with parents. To
carry out these tasks, however, he will need additional specialized training, both practical and theoretical, in social and educational psychology, in their methods of observation and analysis,
and in techniques of guidance.

As to the hardware side, it will be the teacher's function to
direct the various media in the classroom, but apart from this
he will have to give advice in the preparation of course material
and to help curriculum designers by providing them with feedback
information. His experience in the field will be indispensable for
research workers. For these tasks a basic knowledge of educational and social sciences is essential, as is a minimum technical knowledge of media and some training in the handling of
course material.

Independent learning systems will probably be developed outside the school in most cases. However, it will be the classroom,
that is, teachers and students, where learning systems will be
tried out empirically during the development stage. Many learning systems will also be flexible enough to allow for special adap-

tations and adjustments within the school. Furthermore, it is desirable that the teacher prepare his own classroom activities in
a more rational and effective way. To assure students' progress,
therefore, he will have to know more precisely than in the past
what objectives he wants the student to achieve, whether the curriculum or syllabus is adequately designed for certain educational
objectives, and how to assess whether or not the student has
achieved them. Thus, for his active cooperation with research
and development personnel, as well as for his own work with the
students, it is indispensable that the teacher be familiar with the
basic concepts and techniques of curriculum analysis and curriculum development.

It thus emerges that, while freeing the teacher from one function, learner-centered and media-integrated teaching increases
the importance of several others, that is, planning and preparation of learning experiences; observation and evaluation of students' progress; personal guidance of students and parents; management and control of classroom activities; cooperation with research projects. In view of these considerations, the teachers'
fears of loss of professional status and their fear of redundancy
would seem to be little more than the result of wild speculation
and inadequate information. However, it becomes obvious that it
is necessary to revise existing teacher-training curricula and to
develop courses for in-service training.
Social behavior Potentially new learning systems can contribute
to the democratization of learning by bringing about a division of
labor among the various constituents of the teaching-learning system. This will result in a sharing of responsibilities between
teachers and students; that is, it will bring about greater student
participation. This implies that the teacher in the classroom no
longer occupies a monopolistic position that supports autocratic
relationships. Teaching techniques such as team teaching, small
group work, individual tutoring, and others give him more the
position of a primus inter pares and less that of a self-complacent
tyrant. Research has shown that the majority of teachers still
cling to an autocratic behavior pattern and that they find it difficult to adopt democratic leadership. A necessary part of a strategy of implementation, therefore, would be to find those teachers who are least likely to behave autocratically toward the student, and to find them not only among the younger generation, as

these are less likely to hold peer position in the staff. This will
facilitate the integration of new learning systems into the daily
routine of the school. To put the frame factor on social behavior
into a wider perspective, it must be related to the facets of the
interaction process between teacher and student. Thus its vital
importance becomes evident, and it will be necessary to look at
it in more detail from a psychological and pedagogical point of
view.

Concluding Remarks

A reconsideration of the present teaching-learning process and
the introduction of new ways of teaching and learning decisively
alter the role of the teacher. Whereas his present primary function as an information purveyor can be greatly reduced in the future, greater emphasis on the other function is emerging: planning
and preparation of learning situations, personal guidance and
counseling of students and parents, observation and evaluation of
students' progress, management and control of classroom activities, cooperation with research and development institutions.
In order to meet the new functions, teacher-training curricula
will have to be revised and provisions made for in-service training courses. The initial or in-service training courses should
entail the relevant subjects of the following fields:

Social psychology
Educational psychology
Curriculum analysis and curriculum development
Media knowledge and skills
Research methods in social science
Educational planning and institutional management

The introduction of self-contained learning systems calls for
new auxiliary personnel in schools, such as media technicians,
multimedia librarians, and teacher assistants. At present no
proper training systems or career structures exist for these jobs.
Innovations such as those pointed out can be realized only if national authorities consider these changes in the functions of the

teacher in the context of changes in the teacher's social role and
status, and pass appropriate laws.

Selected Bibliography

Bengtsson, Jarl, and Lundgren, Ulf. Modellstudier 1:
Utbildnings-planering och Analyser av Skolsystem.
Pedagogiska Institutionen, Göteborg S, Mölndalsvagen 3. 1968.

Bildung in neuer Sicht: Reihe A Nr. 4 Strukturmodell für die
Lehrerbildung und Lehrerweiterbildung in Baden-Württemberg.
    Reihe A Nr. 8 Aktionsprogramm gegen den Lehrermangel.
Schriftenreich KM Baden-Würtemberg. Neckar-Verlag, Villingen.

Bloom, Benjamin S. "Learning for Mastery," Evaluation
Comment 1, no. 2, 1968.

Bolvin, John O. and Glaser, Robert. "Developmental Aspects
of Individually Prescribed Instruction," Audio-Visual Instruction 13, no. 8, 1968.

Brown, James W., and Norberg, Kenneth. Administering Educational Media. McGraw-Hill, New York-London. 1968.

DeCarlo, Charles. "The 'System' Has Got To Go," College
Management 3, 14-18, 1968.

"Educational Technology," Technical Education and Industrial
Training 10, nos. 11 and 12, 1968.

Flechsig, Karl Heinz. Die Technologische Wendung in der
Didaktik. Universitätsverlag GmbH. Konstanz. 1969.

Glaser, Robert. "The Design of Instruction," in Sixty-fifth
Yearbook of the National Society for the Study of Education,
Part II, The Changing American School. Chicago, Illinois. 1966.
    "Toward a Behavioral Science Base of Instructional Design,"
Teaching Machines and Programmed Learning 2, 1965. Pub-
lished by Department of Audio-Visual Instruction, N.E.A.,
Washington, D.C.

Howson, A. G., and Eraut, Michael R. Continuing Mathematics.
A Proposal for a Systems Approach to the Mathematical Educa-
tion of Sixth Formers Specialising in the Arts and Social or Life
Sciences. Working Paper 2. National Council for Educational
Technology, London. 1969.

Jaquith, Charles E. "An Old School Uses New Tools," Audio-
Visual Instruction 13, no. 10, 1968.

Joyce, Bruce. "The Development of Teaching Strategies,"
Audio-Visual Instruction 13, no. 8, 1968.

Karow, Willi. "Individualisierter Mathematik-Unterricht in
Schweden," Erziehung, nr. 2, 1969.

Loughary, John W. Man-Machine-Systems in Education. Harper
and Row, New York-London. 1966.

McQueen, Mildred. "Are Teachers' Roles Changing?,"
The Educational Digest 24, 1968.

Meiernehry, Wesley C. "Innovation, Education and Media,"
Audio-Visual Communications Review 14, no. 4, 1966.

Müller, Peter, and Thomas, David C. Report on the Interna-
tional Workshop to Prepare an International Conference on the
Changing Role of Teachers Required by Educational Innovations.
Institut für Bildungsforschung in der Max-Planck Gesellschaft,
Berlin. 1968.

O.E.C.D. Educational Technology: A Systematic Approach to
the Teaching/Learning Process. Recommendations to Member
Countries. Paris. 1969.
    Study on Teachers/Études sur les Enseignants. A Series
of Country Case Studies on Training, Recruitment and Utilisa-
tion of Teachers in Primary and Secondary Education. Paris.
1968/69.
    Curriculum Improvement and Educational Development.
Paris. 1966.
    Teaching Physics Today. Some Important Topics. Paris.
1965.

Oettinger, Anthony, with Marks, Sema. Run, Computer, Run.
The Mythology of Educational Innovation. Harvard University
Press, Cambridge, Massachusetts. 1969.

O. I. S. E. The 1930's Multi-Media Kit. A Report on its Use in
Schools. Ontario Institute for Studies in Education, Toronto,
Ontario. 1970.

Postlethwait, Samuel, Telinde, Harvey, and Husband, David.
Plant Science. A Study Guide with an Audio-Tutorial Approach.
Burgess Publishing Company, Minneapolis, Minnesota. 1967.

Rand Corporation. Analyzing the Use of Technology to Upgrade
Education in a Developing Country. Santa Monica, California.
1970.

Razik, T. A. Systems Analysis and Educational Design. UNESCO,
International Institute for Educational Planning. No. 45 of the
Fundamentals of Educational Planning. Lecture Discussion Se-
ries, Paris, 1969.

Research for Better Schools. Individually Prescribed Instruction. Research for Better Schools, Inc., 1700 Market Street,
Philadelphia, Pennsylvania. 1969.

Skinner, B. F. The Technology of Teaching. Appleton-Century-
Crofts, New York. 1968.

Sussmann, Leila, with O'Brien, Marie. Educational Innovation
in the United States. OECD-CERI, Paris. 1969.

Taylor, George, ed. The Teacher as Manager. A Symposium.
National Council for Educational Technology. The Camelot Press
Ltd., London and Southampton.

UNESCO. Stage d'Études sur l'Enseignement Programmé.
Rapport Final. Paris. 1968.

Walbesser, Henry. Constructing Behavioral Objectives. Published by the Bureau of Educational Research and Field Services. University of Maryland, College Park, Maryland. 1968.

Weisberger, Robert, and Rahmlow, H. F. "Individually Managed Learning," Audio-Visual Instruction 13, no. 8, 1968.

Potentially one of the most versatile pieces of educational hardware is the computer. One of the invited speakers was M. W. P. Strandberg (U.S.A.) whose lecture on "Technology in Education" dealt mostly with the use of computers in education. In part he said:

I would like to discuss the computer as a technological aid in education. Again this discussion will be in a restricted sense. I really only want to make an appeal for more creative and imaginative use of the computer in the educational process.

I realize that it will be difficult to find a sympathetic audience to begin with. The very mention of computers in education alienates 99 percent of my colleagues. They feel, and rightly so, that the virtuosity of the professional calculation will be lost by relying on computers. My wife is furious with the computer at the bank that tells her in such an impersonal manner that she has overdrawn our account. I am depressed by the inexorable way the computer extracts my income tax, and the students hate the computer that assigns them numbers and schedules their classes and clamors for the paper work schools seem to require, and so forth.

Whether we like it or not the computer is, and will be, a large factor in our lives, and so I feel we must try to increase its benevolent uses to compensate for its annoyances.

In education this is not an easy task. At first sight the computer appears to be a machine that doles out negative rewards more easily than positive rewards; it seems to be more easily adapted to administering an electric shock for a wrong answer than to handing out candy for a correct one. But this is not so. One may also question whether computer technology in education is of widespread interest. I think it is. Eventually the computer utility, like telephone service, will be realized. Also school administrators, at least in schools that have a special administration, will find they need the computer to do their bookkeeping and management drudgery and they will readily find the money to make one available. Once it is available, the educators will find that time is available on it, and they will put it to their own use.

Let me briefly summarize the uses for computers that have been found in recent studies.

1.

To generate materials to be shown as films, film loops, slides, and so forth—or even to write a bibliographic book, as Sanborn Brown has done.[3]

2.

As a course manager to administer exams, grade them, generate statistical data on exams and students, individualize assignments.

3.

As a tutor (known especially as Computer-Aided Instruction, CAI); this varies from a linear question-and-answer format to an involved Socratic dialogue. The first form could reinforce learning in, say, reading, spelling, or calculus, while the latter should discuss, for example, 4-vector transformations in special relativity.

4.

As a simulator; of electrical circuits, say, or of the very fast, such as a space-time world, or the very small, such as a number of gas molecules coming to thermal equilibrium.

5.

As a calculator; this can have a great impact on science teaching, for the student must learn to use the machine and the course must recognize that the student has this resource.

The impact of any or all of these uses will certainly have to change not only style but content of physics education. Note that these applications are use of the computer as a replacement for some function of the teacher or student. My special appeal in this talk is for the development of uses that are an extension of the abilities of the teacher. For this we really need a new category, number 6.

While I would like to close with enthusiastic support for computers as a teaching aid, I must say something about practical details.

First, developing programs is expensive in time. A single lecture may take a month of programming time to put it into machine language.

----

3. Sanborn C. Brown. Basic Data of Plasma Physics. M.I.T. Press, Cambridge, Mass., 1967.

Second, this technology may seem extravagant, but it really is not compared to some other extravagances we no longer even notice. In one of the large physics lecture halls at M.I.T. we have a hydraulically operated blackboard which I understand cost $5,000 to install. A CRT console needed for use with a central computer does not cost much more than this.

As far as student use of computers goes, for a half-hour of console time per week per student, estimates in 1967 indicate a cost per student per year of about $75. But if one looks at the library budgets of universities in the U.S.A., one finds they amount to $50 to $100 per student; yet we do not even notice the expense or complain that the library is not uniformly used by all students.

Like it or not, the computer is here to stay—let us try to do something beautiful with it!

A lively discussion followed Professor Strandberg's lecture. This was initiated by H. H. Staub (Switzerland), who said:

I agree wholeheartedly with Dr. Strandberg that the computer is here to stay and that we should make the best of it, but I do not think that any of his propositions made here are really good. Let me go through them one by one. "As a generator;" it seems to me that this takes away all the individualism of the teacher. The computer generates teaching programs and teaching aids that are completely uniform, and yet if I remember my own student days, it was just the individuality of the teacher that made life inter- esting. "As a manager;" I consider exams conducted by computer as completely worthless and completely uninformative simply be- cause a computer is a binary device. I can say only yes or no. Good examination questions should be capable of being answered at least ten different ways, and the computer cannot say which is the right one and which is the wrong one. "As a simulator;" again I object very strongly to substituting a computer for actual demon- strations, just as I object to substituting films for simple and cheap demonstrations, as is sometimes done.

I would like to warn very much against the overuse of com- puters, but I am afraid that we are in for some very bad times because of them.

L. R. B. Elton (U.K.): I have a great sympathy for Professor Strandberg although I disagree with him. Perhaps he slightly confused the issues by calling his five uses "replacements." I don't think they are. We all know that the computer is terribly stupid. What we do very well, the computer does badly, and what the computer does well, we do badly, but if we let the computer do the things we do well then we are being stupid. To rebut Professor Staub's criticism—we must not allow the computer to run us, and we must not throw it out just because some people use it badly.

G. Marx (Hungary): The present-day teacher has a twofold job: the job of a machine and the job of a man, of a human being. Let us give the job of a machine to a machine, and let us give more free time to the teacher to make the job of a human being more human. We create a completely wrong impression when we make the students believe that all problems can be solved with simple analytic functions, such as sines and cosines and exponentials. We know in our hearts that all the unsolved problems can be solved analytically. But in practice one finds after leaving the university that one turns to computers to solve the problems. In industry and in science problems will be solved in future by computer, so why not use computers now?

I think we have to use computers to solve problems because nature solves problems in the same way. Computers do not think, so that mathematics and human mathematicians solve problems differently than does nature, since nature does not think either. So that if we want to understand how nature works, we must ask for the help of the computer.

K. Hinst (O.E.C.D.): The computer has been brought into education by industry and we are now trying to adopt it for pedagogical purposes. And there is a great disadvantage to the computer: it is a calculating machine that is used in industry, and we are now trying to use it for educational purposes. This is very difficult, and I think that in order to exploit the potential of the computer in the future we have to investigate how to design special computers for educational purposes. This has not yet been done, but we need to do it.

On the same subject of computer-aided instruction, D. Pescetti (Italy) submitted a paper entitled "Development of Mass Instruction Techniques; Use of the Computer as Teaching Aid:"

During the past decade numerous groups at universities, non-profit institutions, and industrial corporations have begun to explore the possibility of utilizing modern computer technology for education. A widely varying array of such efforts is encompassed by the term "Computer-Aided Instruction" (CAI). Although some recent developments[4,5] in computer-assisted instruction show that large blocks of teaching will eventually be taken over by sophisticated computer systems, this report is primarily concerned with their use as a teaching aid, working in partnership with, but under the control of, the instructor. The computer can be used for processing the educational response data from the students and can thus provide a mechanism for continuous evaluation of student progress and teaching effectiveness.

An achievement test in physics[6,7] is by definition any procedure or device for determining a student's competence in physics. Traditionally, test results are used primarily to assign grades. However, effective teachers find a number of other applications: motivation of students, diagnosis of difficulties, self-appraisal, educational research, and teaching. In the University of Genoa the final examination in general physics for the engineering students (600 students) consists of three parts, a written test (problems), a second written test (30 multiple-choice items), and an oral test. For the past four years the computer has been used for the automatic scanning of the results of multiple-choice tests.[8] The student punches with a pin a special IBM card (port-a-punch), and thus shows the answers he considers right. The punched cards are placed in the computer card reader. The computer then pulls these cards into its electronic scanning device one by one and

----

4. G. Schwarz, O. M. Kromhout, and S. Edwards, "Computer in Physics Instruction," Physics Today, 41, September 1969.
5. D. Alpert and D. L. Bitzer, "Advances in Computer-based Education," Science 167, 1582, 1970.
6. L. Nedelsky. Science Teaching and Testing. Harcourt, Brace & World, Inc., New York, 1965.
7. H. Kruglak, "Resource Letter AT-1 on Achievement Testing," American Journal of Physics 33, no. 4, 1965.
8. D. Pescetti, A. M. Serra, and M. E. Vallauri, "Uso di Test a Risposte Precostituite nelle Prove di Esame di Fisica," Giornale di Fisica VIII, no. 3, 189, 1967.

stores the student's responses in its memory. The correction
and the grading were programmed for and carried out with the
IBM 1620 computer at the Computer Center of Genoa University.
Grading is a less important function of testing than improvement
of teaching and learning. In most cases, tests that are used both
for assigning grades and counseling can also be used for experimental purposes. In the current year tests were administered
continuously during the lecture periods to 150 students of a course
in General Physics II (thermodynamics and electromagnetism).
The computer was used not only for assigning grades but above
all for a detailed analysis of the results: statistics of the frequency of the different answers to each item; correlation between
the responses to different items regarding the same topics; correlation between the answers to items regarding different student's abilities (knowledge; understanding; ability to learn; and
intuitive, disciplined, and imaginative thinking). The discussion
with the students of the statistical results of a test provides a
feedback mechanism both in the learning and teaching processes
and stimulates group work among the students. Once the teacher
knows what his students have or have not learned, he can, if he
wishes, improve his teaching.

The computer time required for the analysis of a test of 30
multiple-choice items (with 5 to 7 responses per item) administered to 100 students is about fifteen minutes with a IBM 1620
computer. In modern high-speed computers this time is greatly
reduced, and more sophisticated applications may be performed,
not only for analyzing multiple-choice but also other kinds of
tests. There are thus wide possible applications of the computer
for direct instructions as well as for assisting the teachers of
large lecture classes. The current university population explosion, the exponential growth of information, and the shortage of
experienced physics teachers accentuate the need of experimenting with methods for mass instruction. Actually these expanding
educational needs have not been matched by increases in the productivity of the educational process. Besides, the cost per student
at all levels is rising. It is necessary to enhance educational
productivity and to enrich the instructional process by the introduction of technology, especially technology of the modern high-
speed computer.

It is important to develop methods that permit effective teaching also in courses with a great number of students. Teaching is a science; therefore valid generalizations can be formulated about it and applied to particular situations. But teaching is also an art and thus depends on the motives, tastes, and talents of the teacher. With computer assistance a teacher can, even in large lecture classes, assess the progress of the individual student in detail; the computer moreover provides the teacher with the means for evaluating his lessons and for measuring the educational effectiveness of his teaching. Since physicists brought the process of measurement to a high level of sophistication, physics teachers should appreciate the need of improved techniques and devices for the measurement of achievement in large classes.

As emphasis to the Conference working group report on "The Technology of Physcis Education" that

today schools are beginning to be equipped with on-line terminals, including in some instances audiovisual displays. In this way, interaction between student and computer becomes an exciting reality,

A. Harashima (Japan) offered a short paper on the "Use of Computer-Generated Animation Films in the Teaching of Physics in Secondary Schools":

There is no doubt that the best way to teach physics in secondary schools is to have pupils conduct experiments themselves or else for the teachers to do lecture demonstrations. However, there are many physical phenomena in which such direct experiments are very difficult or even impossible to conduct.

Among these phenomena we can find, first of all, motions under universal gravitation, if we mean by experiments actual phenomena, not model-like experiments. Fortunately electronic computers, especially those with plotters or display tubes, can provide a good means of generating various kinds of motion with various kinds of initial conditions. Time marks with adjustable equal time difference can be put on the orbit. These marks make it convenient for us to make an animated film of the motion.

To make color films and to combine computer-generated
animation motion with illustrating hand-produced pictures, we
used plotters. Although manual labor was considerable, the motion picture could provide ample information to the pupils. The
list of films which we have made[9] includes (see Appendix D):

Artificial Satellites with Various Initial Conditions
Actual Artificial Satellites
Planetary Motion
Classical Rutherford Scattering
Particles and Waves (photons and electrons)
Earth and Moon Around the Sun

We have found that even physics teachers confess that the
general images of these motions are somewhat different from
those conceived by them. Thus pupils could acquire exact images
of what is going on in these phenomena. Amateur adults can also
understand with interest some of the conclusions of these films.
The education of physics teachers is leading to the creation
of packages of a modular nature which can be fitted into training
courses.

L. R. B. Elton, P. J. Hills, and S. O'Connell (U.K.) had
submitted to the Conference a paper with the title "Self-Teaching
Situations in a University Physics Course," which amplified several points accepted by the Conference from the group report on
"Technology of Physics Education." In connection with the statement just given, these authors wrote:

Tape and Tape/Slide Presentation of Lectures

Audio tapes with or without accompanying slides can be used
either to supplement live lectures or at times to replace them.
A number of investigations have been conducted.
A taped lecture service to supplement live lectures The complete
course on the structure of atoms was recorded during lectures,
using a halter microphone and reel-to-reel tape recorder, and
----
9. A. Harashima, J. Phys. Education Soc. Japan 17, 23, 1969.

transferred to cassette tapes. These were made available to students, together with printed notes that extended the lecture notebook, on a 24-hour loan basis. About 100 tapes were borrowed in the course of six weeks by about one-third of the students, and of these 63 percent expressed a definite liking for them, while none actively disliked them. They were certainly used as an adjunct to lectures, as shown by the fact that attendance at lectures did not drop. Students found them useful for revision, and they liked the more personal style produced by the halter microphone, as opposed to the large, echoing lecture hall. In at least one case, when a lecture was unavoidably missed, the recording provided an easy remedy.

The main problems in providing the service were of an economic kind. They relate to the expense of providing technical staff to do the recording and to transferring it without delay to a number of cassettes, which in themselves can be quite cheap. The scheme also throws an additional burden on the library staff, who issue the tapes and recorders. These difficulties became apparent when the scheme was extended to more courses, and are at present being investigated.

Tapes to replace lectures have been found useful in library instruction, where students can carry a portable recorder to appropriate points in the library, but no attempt has been made to replace normal live lectures. It was obvious that such attempts would be strongly resisted by lecturers.

Replacement of conventional lectures by tape/slide presentations
The situation is different when both tapes and slides are used. As an experiment three of the lectures in the Structure of Atoms course were replaced by tape/slide presentations for the class as a whole. They consisted of edited tape recordings of the lecture material with associated visual material synchronized so that it appeared at the correct points in the presentation. In a questionnaire issued immediately after the lecture, 25 percent preferred the tape-slide presentation and 25 percent the conventional lecture, the rest rating the two methods about the same. But in a more general questionnaire at the end of the course, 45 percent said that they disliked the tape/slide presentation, and only 22 percent that they liked it. The lecturer's view was that the editing had compressed the lecture to the point where it became difficult to follow, and this may well account for the student reaction.

In any case, it is likely that the use of tape/slide presentations with large groups is most suitable for single information
lectures; they have been used with effect in this way for introducing visitors to the university.

Teaching booths The most obvious use of tape/slide presentations
is however in the self-teaching situation. This is facilitated by the
use of teaching booths in which students can work individually at
their own pace. For this purpose a back projection unit has been
developed in which a cassette tape recorder can be used in conjunction with a carrousel slide projector.[10] The booths are situated in the library, which also stores the carrousel magazines
and associated cassettes.

The use of programmed learning in lectures Two investigations
were conducted in which programmed learning with and without
videotape was used in a lecture situation.

Lecture demonstration of an experiment This investigation was
concerned with a means of introducing the class to the periodic
system of the elements by means of practical experiments, in the
absence of a chemistry laboratory. Demonstration of the actual
experiments to the class as a whole was not practicable; the class
was therefore split into six groups. It was then decided to record
the experiments on video tape rather than on slides, since for
practical work motion was judged to be more important than color.
The video tape was used in conjunction with programmed learning
scripts. A pre-test indicated a very wide range of background
knowledge, with those knowing least showing considerable gain
on immediate-recall tests. A fuller analysis of the investigation
is still in progress.

A Comparison Between a Conventional and a Programmed Lecture

Two topics of approximately equal difficulty—polymerization and
metallic bonding—were chosen from the Properties of Materials
course, the first being given as a normal lecture and the second
prepared as a programmed lecture. The latter consisted of material containing a number of questions the students were asked

---

10. P. J. Hills, "Designing a Tape-Slide Unit," The London
Times Educational Supplement, no. 2868, 86, 8 May 1970.

to answer during the course of the lecture by putting their replies
down on a structured answer sheet. The correct answers were
then given immediately before proceeding to the next question.

A preliminary investigation of the results of immediate-
retention tests revealed little difference in the transmission of
knowledge by the two methods, but the results of an attitude questionnaire paralleled almost exactly the figures already given. It
would appear that students have a group feeling about the value of
a lecture which may reflect in terms of student motivation. Informal staff contact furnishes an indication that lecturers like to
lecture, and their enjoyment, which is absent in both the tape/
slide and the programmed presentations, may transfer to the students.

The book is the oldest, and in many ways still the most important tool of Educational Technology. Under the chapter heading of "Duplicated Lecture Notebooks" Elton, Hills, and O'Connell
wrote:

While the Hale Committee[11] was, on not very clear grounds,
opposed to the issue to students of full summaries of lectures,
more balanced opinions[12,13] have been expressed recently and
some positive evidence regarding their value has been presented.[14,15] The issue of printed lecture notebooks to our course
was started in 1964, and over the next few years lecturers pro-
- -

11. Hale Committee. Report of the Committee on University
Teaching Methods. Her Majesty's Stationery Office, London,
1964. Ch. VII.
12. R. M. Beard, Research into Teaching Methods in Higher
Education. Society for Research into Higher Education, London,
1967. P. 25.
13. H. J. Perkin, Innovation in Higher Education—New Universities in the United Kingdom. O.E.C.D., 1969. P. 179.
14. J. McLeish, The Lecture Method, Cambridge Monographs
on Teaching Methods, no. 1, 1965.
15. L. R. B. Elton, The Use of Duplicated Lecture Notes and
Self-tests in University Teaching. National Conference of the
Association for Programmed Learning and Educational Technology. To be published.

vided eight sets of notes, covering in detail the whole course of about 200 lectures in about 700 pages in all. These were issued to the students at 10 shillings per set. The books are printed on the left-hand pages so as to allow students to make their own notes on the right-hand pages.

It is often suggested that the permanency inherent in lecture notes may prevent changes in lectures. It is therefore worth recording that even in the six years since these notes were first produced, they have all undergone substantial alterations. A most important influence in this process has been that for the first time staff have been able to discover the contents of their colleagues' courses without sitting through all the lectures. The resulting discussions have led to many improvements as well as to a substantial integration of different parts of the course without putting an undue strain on friendly relations.

From the start, the reaction of students to this approach was considered very important. It was obtained not only through the normal route of tutorials and formal and informal discussions, but more specifically through a special study[16] and through questionnaires. The first of these was issued to students in the Mechanics of Particles course in 1967. The replies showed that all students owned the notebooks and nearly all found them very helpful, with the details being just about right. On the whole, students rarely read ahead and they stated that they could at most have learned only part of the course without attending lectures. The vast majority rarely missed a lecture and most took additional notes at times. Answers to open-ended questions revealed that many students found the mathematics difficult and would have liked more detail and more physical explanation of the mathematics. They overwhelmingly felt that the scheme should be extended. They were prepared to pay accordingly, and they listed among its advantages that it enabled them to follow the lectures without having "to get everything down," to catch up on a missed lecture, to read in advance, and to revise better. The only dissenting voices came from two students who found that note taking kept them more awake. A very similar questionnaire, yielding almost identical results, was given to students in the Structure of Materials course

----

16. S. O'Connell, "From School to University," Universities Quarterly 24, 177, 1970.

in 1970. This course was chosen because it is the most descriptive, while the Mechanics course is more mathematical. In both courses, attendance was checked and rarely dropped below about 70 percent of maximum. Little is known about what proportion of a possible maximum normally attends lectures, but even where these are compulsory, as they are not at Surrey, a 75-percent attendance is considered satisfactory.[11]

The existence of lecture notebooks made it possible to discover the extent of previous knowledge that lecturers expected the students to possess. As a result, a number of preknowledge surveys in different subjects were constructed, which revealed consistent and often previously unsuspected gaps in the students' knowledge.[17] This has led to a modification of the lecture courses and also to a proposal to construct a number of self-teaching modules, to help students whose preparation was inadequate.

The situation where students lack the necessary preknowledge does not of course arise only at the beginning of the students' university career; this was made apparent by an experiment in a second-term course, designed to orient students to the subject matter of a lecture prior to the lecture. Often the introduction of a topic in a lecture is the first time that it has been considered by the student. In this experiment he was required first to establish his own state of knowledge by means of a diagnostic test and then to take steps to correct deficiencies. He was then taken through a set of structured notes that included important words or phrases keyed into remedial material. Finally, short lectures were given which served mainly to troubleshoot problem areas and to expand points which were felt by the lecturer to be particularly relevant or difficult.

The experiment, which covered the topics of ionic bonding and Van der Waals forces in the Structure of Molecules course, extended over three normal lecture periods of 50 minutes each, during which the lecturer gave two short lectures of 15 to 20 minutes each. The average score on a pretest for the 112 students taking part was 44 percent, and on an immediate post-test it was 86 percent. Of 72 students who answered a questionnaire on the

----

17. S. O'Connell, A. W. Wilson, and L. R. B. Elton. "A Preknowledge Survey for University Science Entrants," Nature 222, 526, 1969.

experiment, almost exactly half preferred the system to conventional lectures and thought that it led to better remembering and understanding, while just under one-half found no difference. Fewer than 10 percent preferred the conventional approach.

One of the difficulties that students face when they first come to university is that they are very uncertain whether they are making adequate progress, but at the same time they do not wish to reveal their possible ignorance to their teachers. To overcome this problem, a book of self-tests was devised to accompany the lecture course on Mechanics of Particles (Elton, 1970). The book consists of four parts:

1.
An introduction, with a flow diagram, giving instructions on how to use the book.

2.
A set of 100 multiple-choice questions arranged in sections corresponding to the sections of the lecture notebook. These are to be attempted immediately on completion of each section in the course.

3.
A code number attached to each question leading to the correct answer in the next section. The answers are scrambled so as to prevent students from accidentally seeing the answer to the next question.

4.
Similarly, a code number in the answers leading to an explanation in the final section. This is designed to make sure that the student obtains the right answer for the right reason.

The book thus serves not only as a testing device but should also aid learning. How far it has in fact led to improved learning is difficult to establish, because it was not possible to use a control group. Comparison with the results obtained in years prior to the introduction of the self-tests was also impossible, since the latter coincided with the move of the University from London to Guildford, an event that masked all other changes. Results of a questionnaire showed that 42 percent of the students used the tests "frequently" or "always," and only 7 percent did not use them at all.

The following references are also pertinent to the subject of issuing lecture summaries to students:

Boud, D. J. The PEMS Physics Laboratory, a Pilot Enquiry. University of Surrey, unpublished. 1970.

Elton, L. R. B. "New Courses at the University of Surrey," Physics Education 1, 89, 1966.

Kilty, J. M. Submitted to Visual Education for publication.

Young, M. Innovation and Research in Education. Routledge and Kegan Paul, London. 1965. P. 8.

As an illustration of resources for Educational Technology in the U.S.S.R. and in Bulgaria, a list was compiled for these two countries by E. A. Shekuteva (U.S.S.R.) and V. Metev (Bulgaria):

1. Educational Material (Mainly Software)	2. Aids for Technical Education	3. Audiovisual Equipment (Hardware)
1.1. Wall Charts Photographs Maps .	2.1. Full-Size Working Machines	3.1. Epidiascope Overhead Projector Stereo-Optical Systems Slide Projector
1.2. Biological Models Geographical Models Chemical Models	2.2. Machine Parts Machine Sections	3.2. Film Projector Film-Loop Projector
1.3. Educational Toys	2.3. Small-Scale Working Models	3.3. Radio, Earphones Gramophone Tape Recorder
1.4. Mechanics Kits Electrical Kits Electronic Kits Automation-Control Kits	2.4. Two-Dimensional Dynamic Models	3.4. Television Video Recorder
1.5. Transparencies Slides	2.5. Simulation Models	3.5. Computational Devices Computers

1.6.
Films
Film Loops
1.7.
Measuring Instru-
ments for Sciences
Mathematics
1.8.
Magnetic
Video Tapes
Records
1.9.
Language Labora-
tory Physics Ma-
terials
Charts

3.6.
Language Labora-
tories

The Conference group discussing The Technology of Physics
Education found it useful to compile a glossary of technical terms
used in their discussions. To aid in clarifying their report, this
glossary is appended:

Communication Center:
The place where teachers retrieve or exchange information, materials (such as books, audiovisual aids, packages, tapes, charts,
and so on). Commonly it would be situated at teacher-training institutions. The term refers to a new version of the traditional
library. As such it can also exist at schools where it is associated with the Resource Center (see below).
Correspondence Course:
A course that is administered by sending the learning material
to the student by mail. The interaction processes between teacher and learner are taking place individually through correspondence and not in face-to-face contact.
Hardware and Software:
The term hardware comprises all kinds of media, that is, material and means used to convey information (for example, the radio or television set, the film projector, computer).
In contrast, the term software refers to the program or the
message or information which is transmitted through it. This in-

cludes such things as for example the picture that is put on a transparency for an overhead projector or the sound on a tape.
Mass Media:
This pertains to radio, television, films, and printed material, through which a multitude of people can be reached at the same time.
Module:
A basic standardized unit. In a curriculum or a learning package (see below) it characterizes the various parts of which a course consists and which for design purposes can be looked upon as a whole.
Package:
A prefabricated course consisting of a multiplicity of media methods and procedures in an integrated manner. Packages can be produced for use by the teacher in the classroom and/or for individual use by the student independent of the constraints of time and space. A package may consist, for example, of a teacher's guidebook, textbooks, working sheets, film loops, audio tapes, transparencies, experimental equipment, test sheets.
Resource Center:
A material and data bank (see also Communication Center) that includes workshop facilities for the production of equipment.
Systems Approach:
In Educational Technology the term "system" refers to the interdependent and hierarchical structure or organization of the technical and/or social elements, respectively, of individuals. Systems approach then signifies a methodology of rational analyses of a system or its elements that makes it possible to decide on the best sequence of steps to be taken for achieving set objectives.

Chapter 9

Special Problems of Developing Countries

The report of the working group on Special Problems of Developing Countries[1] was given to the Conference as follows:

First, the group identified criteria useful in categorizing the
so-called developing countries from the physics-teaching viewpoint. Next, it identified ten problems that are peculiar to or at
least especially acute in countries described by these criteria.[2]
Finally, it discussed these problems in detail and, where possible, listed guidelines for action toward possible solution.

The Criteria

1.
The prevailing technological environment gives the preadolescent
child little opportunity to be surrounded by the paraphernalia that
generate a certain degree of technological experience.
2.
The per capita income is below the level required for self-suf-
ficiency. It is pertinent to note that the per capita income among
the countries of the world ranges from $ 40 to $ 400 per year.
The percentage range of per capita income within the developing
countries is often greater than in the advanced countries.
3.
Although education is far from universal, the rate of increase in
the school-going population is greater than that of the production
of teachers.

----

1. Members of the working group which discussed this subject
were: A. V. Baez (U.S.A.), Chairman, W. A. Blanpied (India),
Rapporteur, R. B. da Costa (Brazil), A. Harashima (Japan),
N. Joel (UNESCO), J. A. O. Sofolahan (Nigeria).
2. See also P. G. de Paula Leite, "Observations on the Teaching
of Physics in a Developing Country," published in Why Teach
Physics? M.I.T. Press, Cambridge, Mass., 1964. P. 11.

The Problems

1.
Language
2.
Transfer of curricula and methodologies from the developed countries.
3.
Creation of indigenous methodologies.
4.
Severe limitations in human and material resources and in technical know-how. The wide gap between resources available in the principal cities and the hinterland.
5.
Indigenous concepts and traditional beliefs that often militate against the teaching of scientific methodology.
6.
Wastage of trained teachers.
7.
Low status of teachers.
8.
Poor opportunities for teacher retraining.
9.
Poor diffusion of information.
10.
Insufficient international assistance.

Guidelines

1.
Language In many developing countries a large number of local languages and dialects are spoken, while the educated élite is conversant with an official or semiofficial international language. Usually the study of this language is compulsory at the middle-school level. Since textbooks and indeed the requisite scientific vocabulary usually do not exist in local languages, secondary-school students learn the sciences in their second or third language, in which their fluency may be poor.
　　Although the problem assumes different forms in different places and is likely to change with time, the group recommends the following:

1.
Where practical and possible, teachers should be encouraged to
develop learning materials in the local languages as well as in
the official language.
2.
The language used by teachers in discussing physics must be
comprehensible to his students. Thus in their training, teachers
must be made fully aware of the communication difficulties involved. Some form of specialized language training aimed at this
is recommended.
2.
Transfer and/or creation of indigenous methodology Although
physics as a science has universal applicability, teaching methodologies do not. Rather, they are strongly dependent on the conditions prevailing at a particular time and in a particular place.

For example, during the last decade several attempts were
made to introduce the PSSC[3] physics into the developing countries.
For a number of reasons the course was found to be unsatisfac-
tory. (The particular examples of Latin America and India[4] were
cited.)

A physics course consists of more than a text. Attempts to
use a course like PSSC without its associate laboratory and demonstration apparatus were often made, with poor results. More
important, perhaps, is that the aims of a course developed by
curriculum-reform groups in an advanced country are rarely
relevant to the needs and conditions prevailing in a developing
society. The background common to students in advanced countries which is assumed for such courses does not exist in the developing countries. Nevertheless, the exposure of teachers in developing countries to the product of curriculum-reform groups in
the advanced countries has had a historic effect which must be regarded as beneficial. Contact with these courses generated considerable discontent with existing methodologies and stimulated
the desire for curriculum reform. Thus, as it became clear that
new courses from the advanced countries could not serve the needs

----

3. See Appendix C.
4. Paper contributed by W. A. Blanpied and V. S. Nigama: "A
Note on the Indo-U.S. Summer Institute Programme for Secondary School Physics Teachers."

of the developing countries, it also became clear that indigenous courses would have to be developed.

In the light of the discussion the group recommends that teachers in developing countries be encouraged and aided to work out materials relevant to local needs and conditions. Indeed, projects in curriculum structure might well be part of the pre-service training of teachers. In both their pre- and in-service training, teachers should be exposed to courses developed by curriculum-reform groups in the advanced countries. However, these should serve only as guidelines for developing indigenous methodologies.

In-service workshops in curriculum development are highly desirable, since the teachers themselves are involved in the education-reform process. It has been noted that such workshops also increase the self-confidence and self-respect of the teachers.

Although the creation of physics courses must be an indigenous effort, foreign experts, like foreign courses, can and have had positive catalytic effects.

3.

Limited human and economic resources Because of the limited economic resources available, few schools are able to afford the type of apparatus associated with contemporary physics courses in the advanced countries. Furthermore, because of their long working hours, teachers are generally hard pressed to find time for improvisation. Thus the curriculum in a developing country must be adapted to take advantage of resources available in the environment. Full advantage should be taken of astrophysical, meteorological, and geophysical phenomena, for example. The prevalence of traditional animal- and man-powered machines should suggest a wealth of applications to physics teaching.

In their training, teachers must be educated to use such resources and trained and encouraged in improvisation of simple apparatus. They should also be trained in utilizing their students in improvising apparatus, since the aid of students will not only amplify their own efforts but in addition can be a valuable part of the students' education.

Ultimately, science teaching in the developing countries can be improved only if more apparatus is made available. Thus the various governments must provide more material resources. In addition, customs duties and foreign-exchange policies should be

adjusted to the benefit of science teaching, on the one hand permitting the importation of necessary apparatus, while on the other
encouraging indigenous manufacture where such manufacture is
feasible.

The group recommends strongly that well-staffed centers
for the improvement of science teaching be established or, where
these already exist, that they be strengthened. Such centers can
serve a variety of functions. In particular, there can be apparatus banks, as well as institutions at which teachers can come to
be trained in the improvisation of apparatus. They can serve as
obvious centers for curriculum development projects. Additional
functions of the centers are discussed later in this report.[5]

4.
<u>Indigenous concepts and beliefs</u> Frequently superstitions, tradi-
tional beliefs, and cultural patterns are inconsistent with scientific methodology. Thus, in order to convince the student of a
physical principle, the teacher must use demonstrations rather
than dogmatic assertions, but he certainly must not ridicule the
students' cultural upbringing.

Physics teaching in the developing countries has traditionally emphasized fact rather than process. Teachers must be trained
to involve their students in situations in which they can be led to
practice and discuss the scientific method for themselves. The
use of open-ended laboratories can serve this purpose, for example.

5.
<u>Wastage of trained teachers</u> The group noted that in many devel-
oping countries trained teachers frequently leave the profession
as soon as an opportunity is found for employment in a position
offering a better salary, higher status, and better opportunity
for advancement. This rapid turnover of teachers is especially
alarming in societies that require increasingly large numbers of
trained teachers. While the problem is serious, it is inherent in
the nature of developing societies and cannot be solved by a reso-

----

5. See Annex I, p. 210. For a discussion of the role of one possible
set of centers, see <u>Physics in India: Challenges and Opportunities</u>,
Chapter 6. Available from R. D. Deshpande, National Council for
Science Education, 9 Ring Road, Lajpat Nagar IV, New Delhi 24,
India.

lution of this Conference. It is possible that a training that gives
the teacher some pride in his profession and the opportunity to
be creative might effect some small reduction in wastage. Giving
publicity to outstanding teachers is one method of increasing the
requisite professional pride.

6.

Low status of teachers Since the status of secondary-school teachers in developing countries is not high, an insufficient number of
qualified people go into and stay in the profession (see item 5).
Again little can be done about the problem through teacher education. Ministries of education in the various countries should give
due consideration to improving the salaries, working conditions,
and advancement opportunities of teachers.

7.

Poor opportunities for upgrading teaching ability Typically the
secondary-school physics teacher is isolated from his colleagues,
has had little contact with developments in physics and physics
teaching since receiving his certificate, and often has had only
two or three years of schooling beyond the level he is expected
to teach. Thus, in-service training is even more important in the
developing than in the advanced countries. Given the isolation of
most teachers, it is desirable that much of this training be brought
to the teacher and that it be continuous.

Centers for the improvement of science teaching can play a
primary role in such training by arranging in-service workshops
for teachers. Consultants attached to the centers should go into
the schools in the hinterland to work with local teachers, helping
them upgrade their ability.

The feasibility of using such self-retraining methods as correspondence courses and educational television should be investigated by the centers. The centers are also the obvious places for
implementing these methods.

8.

Diffusion of information The loneliness and isolation of the typical
teacher in the developing societies has already been noted. A number of steps should be taken to enable him to keep in touch with
developments elsewhere in the country and in the rest of the world.
A continuous assistance program carried out by a center for the
improvement of science teaching has been mentioned under item 7.
The formation of science teachers' organizations with strong local

branches should be encouraged. Teachers in their training should
be taught to consult journals with information pertinent to physics teaching, and these should be made available in their school
libraries. Science-teaching improvement centers, possibly in cooperation with the teachers' organizations, can publish and disseminate newsletters to all physics teachers in the country. The
centers should subscribe to international publications, such as
UNESCO's New Trends in the Teaching of Physics, and take steps
to diffuse the pertinent information throughout the country, possibly by means of the newsletter.
9.
Inadequate international assistance International assistance and
cooperation have played a large part in improving the educational
system of the developing countries. The growth of better science
teaching will require large expenditures, which will in many cases
have to be provided by these same means. Among the organizations that can help the developing countries are the specialized
U.N. agencies, the several international cooperation agencies involving groups of donor countries, and private charitable foundations. Most donor organizations welcome requests based on well-
thought-out projects documented by adequate preliminary investigation or experience.

We would urge countries to attempt to solve educational problems by using their own resources. Many of the suggestions put
forward in this report can be implemented by existing institutions,
using personnel that is already available. In most developing countries small groups of imaginative and creative teachers exist who
can undertake curriculum reform, in-service improvement seminars, or even establish small-scale centers for providing the science teachers with the help they need. Universities and teachers'
colleges in the developing countries can accept the responsibility
for managing projects designed to improve science teaching.

Cooperation among developing countries may be a useful mechanism for solving common problems and reducing costs.

Programs that are supported by international or bilateral cooperation must respond to real needs, and must be shaped so as
to match indigenous conditions. Such programs often provide for
assistance in the form of cash grants, fellowships, equipment,
and the services of experts. Of these, the most sensitive component is the last one. People who undertake international assign-

ments must be flexible enough to adjust themselves to their new temporary assignment. Recruitment agencies should not rely on pre-mission briefing to remove the prejudices and inflexible opinions that an expert may possess. Rather, selection committees should include a competent member from the developing country who can judge the attitude of the proposed expert.

Programs operated through international cooperation should stress more than heretofore the training of local personnel through fellowships, so that the use of experts need not be perpetuated beyond a brief period of project initiation.

Annex I

Centers for the improvement of science teaching can serve one or more of the following functions (restricted in some cases to physics alone or generalized in others to the whole school curriculum):
1.
Provide information on recent advances in science education, especially on new approaches and new materials for science teaching.
2.
Have available materials (books, apparatus, kits, films, other aids) produced by recent national and international science-curriculum projects, which can be consulted and tried out by science teachers and other interested persons.
3.
Have workshops, laboratories, and other facilities, where science teachers can develop and produce prototypes of apparatus and kits of their own design.
4.
Establish a meeting place for science teachers to discuss questions related to their work, both among themselves and with scientists, psychologists, media specialists, and so on.
5.
Initiate experimental projects for the design of new courses and the preparation of the corresponding new learning materials.
6.
Collaborate with the educational authorities toward the effective implementation of new science curricula in schools.

7.
Experiment with new approaches in evaluating student achievement.
8.
Provide a variety of opportunities for teachers to become more self-reliant and more familiar with the new approaches and the new learning materials—from very informal gatherings organized by the teachers themselves to formal courses of varying types and duration.
9.
Provide facilities for extracurricular science activities, such as science clubs, and so forth.

This list is not exhaustive, nor is it suggested that a single center would serve all these functions. Ideally, a center of this type should start as a local initiative, as a small-scale experimental project. For instance, the physics teachers of, say, five to ten schools could take the initial steps; or the initiative could come from a department of a Ministry of Education, from a university or teachers' college, and facilities provided by that university, a college, or a school. In time, as the center develops, it may increase the size and scope of its activities and become a national institution with activities all over the country. At this stage, it may receive international assistance. However, great care should be exercised in ensuring that the new science curricula respond to the real local needs. This means that in the developing countries ongoing projects should receive support. Other potential innovators must first be identified. They should then first be given opportunities to develop their capabilities and may then receive support for their work.

This report of the working group of the Conference on the Special Problems of the Developing Countries drew largely on the work of a number of conferees with extensive experience in these countries. The guidelines suggested for the handling of the problem of language referred to two contributed papers, one by J. T. Macfarlane (Rwanda) and the other by R. L. Krans (Netherlands). This latter paper has been summarized in Chapter 2 (see p. 3). In considering the special problems of the secondary-school physics teacher in developing countries, Macfarlane writes:

In most African countries, teaching in secondary school is carried out in a foreign language. Even in the senior part of the school, the pupil's command of the language of instruction is weak, and in many cases so also might be the teacher's. Whether it is a question of understanding a class discussion or of reading a textbook, the student in many cases is at a considerable disadvantage. He inevitably resorts to memorization. By this technique he might indeed come to know a certain number of the facts of physics or at least be able to repeat them on request.

This problem will be with us for a long time, and indeed perhaps forever in the less technically advanced countries. The older bilingual or multilingual countries have met the same difficulty, and in looking for optimum ways of alleviating the situation the solutions should be studied and in some cases adopted. The production of teaching materials in vernacular languages is not progressing quickly, nor is it indeed thought by all to be the proper solution for science teaching at the secondary level. The élite of a developing country must learn an international language thoroughly; its use as a medium of instruction is a necessary part of the learning process.

In the context of Southeast Asia, this linguistic problem has been explored at great length in a report by Richard Noss.[6] One conclusion is that if the student is to be able to study and work in a technical field, an intimate knowledge with an international language is essential, and this can be assured only if instruction at the secondary level is in that language. It has been shown that real fluency in a language is seldom acquired if its use is delayed beyond the age of fifteen.

On the other hand, primary schooling, directed at an increasing proportion of the population, must be carried out in a national language. Thus the pupils' knowledge of the international language, at secondary level, will be incomplete. The secondary school teacher of physics must therefore accept this state of affairs and use his classes for the double purpose of teaching the subject and strengthening the students' use of the technical vocabulary in the language of instruction.

----

6. R. Noss. "Linguistic Policy," in Higher Education and Development in South-East Asia. Vol. 3, part 2. UNESCO and IAU, Paris, 1969.

Considering a further point discussed in item 8 of the working-group report concerning the beneficial effects of a visit from a consultant to a school, Macfarlane says:

The teacher who works with the substandard facilities of a secondary school in a poor school system needs constant encouragement and advice. Most education departments provide short courses during which teachers in service learn new techniques, or even just refresh their knowledge of the subject they are teaching. Such occasions are of tremendous value to the participants, since among other things they become aware that they are not alone in facing insuperable difficulties.

Very often the beneficial effects of this kind of effort are short lasting, because the stimulus disappears as soon as the teacher is back in his usual surroundings. The missing element is one of regular intervention.

All secondary-school systems have an inspector service. What is needed in the outlying areas is a monthly visit from an experienced teacher who can help in a multitude of ways to render the situation more palatable, especially for the pupils. One of the inspector's activities should be actual teaching of some classes. This will serve as a demonstration lesson and also be a highlight of the month for the pupils. The visitor can help the teacher with minor apparatus repairs or show him how to use unfamiliar equipment that might have been obtained from an apparatus-loan service. The teaching plans for the following month can be prepared, and any expected difficulties can be foreseen and so avoided.

This way of operating the inspection service does imply an increased manpower investment. In most environments where travel is difficult, monthly visits would require an additional experienced teacher for each twelve of fifteen schools. If the system were to be extended to all subjects in the school the budget requirements would be prohibitive. It should be possible however to carry out such a scheme in all the sciences and in mathematics if the usual inspection activities are curtailed or carried out only as subsidiary to this consulting service. Often the same person could be the visitor for two or three subjects.

While most education departments jealously retain control over their inspection service, they might well look with favor on the organization of a consulting service, such as that outlined here, by a teacher-training institute or the university. The inspectors' association with a teaching organization would give them easier access to facilities they would need for carrying out the program. A given person might act as inspector only in occasional years, being at other times a member of the institute's teaching staff. Finally, the consulting service might be coordinated with an apparatus-loan service, providing an integrated system of assistance for the secondary-school science teacher.

Discussing the now extensive experience which exists concerning the interaction of a new program derived from a developed country with a developing country, W. A. Blanpied and V. S. Nigama (India) confirm the view expressed by E. M. Rogers (U.K.) in his paper on the Nuffield Project. Rogers points out that

"since this is a course arranged to fit the conditions that obtain in England, its materials are not likely to be of direct use for teaching elsewhere. Yet the Teachers' Guides offer suggestions and commentary that should make them of considerable value in any library for teachers in other countries."

In "A Note on the Indo-U.S. Summer Institute Programme for Secondary School Physics Teachers," W. A. Blanpied (U.S.A.) and V. S. Nigama (India) say:

Whereas the institute participants in these early years regarded a program based upon the compact PSSC course as interesting and perhaps rather pleasant, they were able to make little direct use of the curriculum in their day-to-day work, primarily because PSSC physics did not conform to the syllabi used in their home institutions. Indeed, it is questionable whether PSSC physics or any other introductory foreign curriculum in unadapted form would be suitable for Indian secondary students. Of necessity such courses make liberal use of experiences they regard as common to all students to stress the basic concepts. But examples based on jet aircraft behavior or the experience of an ice skater, for instance, are virtually meaningless to Indian students.

In fact, it can be argued that any attempt to impose a uniform curriculum over the whole of a country as large and diverse as India must of necessity be futile. The Indian constitution places control of education in the hands of the states, leaving the center with the power to advise and to aid them financially in their educational programs. In addition, instruction in the overwhelming majority of the secondary schools is carried out in one of the fourteen local official languages recognized by the Indian constitution, or in some cases even in a local dialect.

Thus after the first few years it became evident that if the institutes were to be relevant to Indian physics education, the major responsibility for planning them would have to be placed in the hands of the local directors who were expected to be more in tune with local conditions than any central authority no matter how well intentioned.

Of particular interest was the following information from a paper contributed by E. B. Kvashin (Nigeria) on "The Training of Qualified Physics Teachers for Secondary Schools in Nigeria":

In Nigeria, general education is free and is compulsory for six years. The system of secondary schools and higher education has grown considerably as a result of this policy. There are now five universities and a number of technical, medical, and economics schools and pedagogical institutes in the country. Because of the rapid changes of all kinds in Nigeria in the past ten years, there has been no opportunity to educate enough secondary-school teachers, especially teachers of science and mathematics. Fifty-nine percent of the teachers are Nigerians; 41 percent have been invited from other countries; and only 50 percent have had more than one or two years of experience in teaching. Also, the percentage of applicants qualified to enter the pedagogical institutes is low. At Adeymi College of Education only 150 to 180 are admitted out of the 3,000 who apply, mainly because most cannot reach the required minimum grade of 40 percent in the entrance examination.

Among the reasons for the low standard of science education in physics are the following:
1.
There are not enough good physics teachers in the schools now. They do not have the necessary experience in physics teaching, and they are overloaded.

2.
Many schools are very poorly equipped. The students have no opportunity to get any laboratory experience and skills.
3.
The textbooks are too old. They give only the main scientific laws,
and do not help in understanding science.
4.
New methods of physics teaching have not spread over the country.
Teachers teach mainly blackboard physics. Only 11 percent of the
teachers have had the special education needed to prepare demonstrations. The teacher talks, the students listen and repeat what
he says, but logical thinking does not develop.
5.
Students coming from different secondary schools have had very
different science instruction; although they have had several years
of general science, they have had physics only during the last two
or three years.

Strong action is needed to change teacher education. It has
begun with the establishment of six teacher education institutions,
planned with the help of UNESCO. The instruction and living facilities in these schools are free for all students. At present, 80
percent of the students in postsecondary institutions are future
teachers. Also, in 1968 the Science Teachers Association of
Nigeria introduced a new system to raise the level of secondary-
school teacher education by setting up several special schools for
performing educational experiments. They provide basically a
three-year program, in which the student must study the English
language, pedagogy, and two subjects. They choose a combination, such as two sciences or a science and mathematics. Future
physics teachers study physics for nine semesters, including one
semester of calculus at the beginning and a semester of practice
teaching in a school before their course is completed.
Teacher training in Nigeria has developed with the country.
We need highly qualified, well-trained native teachers for the education of our young people. The type of pedagogical institution discussed here is in our minds the most practical form under present
conditions. This sort of institution provides a place for developing
programs and introducing new methods of physics teaching. Secondary-school physics teachers will be further trained in these
institutions in periodic retraining courses and evening sessions.

In an invited paper, N. Joel (UNESCO) provided an outline answer to the question "How Can International Organizations Help Physics Teacher?"

I have been asked to speak not only about UNESCO but also about other international organizations. For this purpose I have interpreted the word "international" in a rather flexible way. That is, I will be referring not only to what are strictly speaking international organizations, that is, those that are open to countries in all regions of the world—of which there are not many interested in helping teachers of physics. I will refer also to what we call regional organizations, which include a number of countries in a limited region of the world. I shall be giving several examples of both; these, I hope, will show that there are many different possible ways of working internationally for the improvement of the teaching of physics, in both government and nongovernmental organizations, and with varying degrees of internationalization.

Furthermore, in discussing what can be done by international organizations to help physics teachers, I have understood this to mean that we should first see what they are already doing and then discuss what could be done if there were more means available to do so. I will be careful, of course, to separate these two very clearly, so as not to confuse what is real with what is desirable.

But before we discuss specific examples, let me briefly refer to some of the main ingredients that go into a program aimed at improving physics education in a country. And I would like to make a distinction between those ingredients that are relatively cheap and those that require a lot of money, time, and skill.

An ingredient that falls into the first of these categories is information. It happens to be one of the first elements required if we think in terms of the timing of the various steps. This includes information on what is going on in the country, plus a clear statement of the needs. But I am mainly thinking here of information on what is being done in other places to solve similar problems: information on new physics courses, on new teacher-training programs, on the design of new learning situations, on low-cost apparatus, on the effective use of mass media, on ways of letting every student perform to the best of his ability, and so on. This kind of information, on how science education gets improved in other places, is needed as much in the advanced countries as in the developing ones. In both it can help those individuals and in-

stitutions that have already decided to do something, but it can
also have a strong stimulating and catalytic effect in those places
where the need for change is not yet widely recognized.

It thus seems to me that the international exchange of information and ideas, through publications, meetings, seminars, and
other means, has high priority. But, having whetted the appetite
of those who want to innovate, we must do something more to help.
Here is where the other ingredients come in—the expensive ones.
They include trained personnel, money, institutional structures.
They also involve assisting the developing countries in the preparation of their long-term plans and in the mapping out of strategies for action.

Let us now see how the international organizations face this
challenge in the field of physics education. And quite naturally,
I propose to begin with the International Commission on Physics
Education of IUPAP, as this is the commission which the world's
physics community has set up to deal with the educational aspects
of physics. The Commission's sustained record in the international field throughout the last ten years or so deserves a detailed
report. I shall make it brief, however, because I assume that
most of you are familiar with it.

International Commission on Physics Education

The Commission's activity has been mainly concerned with international meetings and publications. There have first been the three
International Conferences on Physics Education, held in 1960
(Paris), 1963 (Rio de Janeiro), and 1965 (London), each with a
specific theme. Then, two international seminars: one on the Education of Physicists for Work in Industry (Eindhoven) in 1968 and
one on the Role of the History of Physics in Physics Education
(Cambrige, U.S.A.) in 1970. Each of them has led to the publication of a report, all of which been very well received by those interested in the advancement of physics education.

The Commission has also been engaged in a number of writing
projects in collaboration with UNESCO. One of these was the Survey of the Teaching of Physics in Universities. This Survey covers
six countries and, although written around 1963 and published in
1966, it is still in demand. It gives much useful information, even
though some of it has inevitably been partly superseded by recent

university reforms. Then, there are the series on "New Trends in the Teaching of Physics" — of which volume II is about to be published — and the Source Book for the Teaching of Physics in Secondary Schools, also in preparation at present. For both of these publications, the Commission is serving in an advisory capacity to UNESCO.

We can now ask ourselves: Is the International Commission on Physics Education working in the right direction? I think the answer is definitely yes, as follows from my earlier remarks concerning the great need for a global information program mainly through publications and meetings. Is it doing enough? The answer is: however much gets done, it will never be enough to meet all the needs. But, considering the very limited means at its disposal, I think the Commission is doing much and is doing it well. This may be an appropriate place to express UNESCO's appreciation for the work of the Commission.

UNESCO

We must now turn our attention to the activities of UNESCO in physics education.

One of our experiments on new approaches has been the UNESCO Pilot Project on the Teaching of Physics. Although this was started some time ago — in fact, its first phase took place in 1964 — I have at least two reasons for referring to it now: first, because several of our present activities in physics are a followup of this Project and are being carried out in the light of the experience gained through it; and second, because this Project did constitute a rather interesting way — and I think also a rather effective way — of helping physics teachers. During its first phase of one year, which took place in Brazil, 25 physics teachers from 8 Latin-American countries worked under the guidance of specialists in physics, physics education, psychology of learning, and film production provided by UNESCO. One of the aims of this one-year exercise was to give these 25 teachers an opportunity to receive training in the design, development, evaluation, and production of part of a physics course; they were therefore given the task of doing this for a course on the physics of light at the secondary level. By the end of the year they had designed, developed, tested, and produced a first version of a set of laboratory kits, programmed instruction manuals to go with the kits, TV programs, films, and a teachers' guide. Another purpose of this project was to see whether this operation could be done in one of the developing countries. The project proved—as are our similar projects in Asia (for chemistry), in Africa (for biology), in the Arab States (for mathematics), and in several regions (for integrated science)—that this is perfectly feasible. In the meantime it has become obvious that this is also necessary; new courses and new learning materials should, whenever possible, be prepared, tested, and produced locally, with the active participation of local teachers and science educators. This active participation of local personnel has manifold advantages, apart from increasing the suitability of the learning materials to local conditions. I will mention here only two of these additional advantages: on the one hand, those of the participants who contributed most have in the meantime been able to become leaders in science-education reform in their countries; and on the other hand, even those who contributed least have learned to use the new approaches and modern techniques, effectively have become more open to innovations and have increased their self-reliance.

We now take the view that science-education reform has to be set up as a continuous self-correcting process and has to be approached in a more comprehensive manner, attacking on many fronts at the same time. Nonetheless, while such a systems approach would yield excellent results in those places that are ready for it, it seems to me that there are also other places where the interest in science-education reform has yet to be strengthened and where potential innovators have yet to be identified and trained by means of less ambitious experimental projects. Such pilot-scale projects constitute in many cases a necessary first stage that opens the way for a later stage of full-scale science-education reform.

As a follow-up to that one-year phase of the physics teaching project, UNESCO sponsored a series of regional seminars for physics teachers, at each of which 30 to 40 teachers from 5 to 15 countries had a chance to work for a month with the ideas and materials developed by the UNESCO Project as well as with those of the most important contemporary physics-curriculum reform groups, and in some cases to discuss them with their main authors. Eight such seminars have taken place so far and with two exceptions there have been at least three different curriculum groups represented on the staff. In this way we are trying to avoid giving the impression that there is one best way of teaching physics—and that we know which one it is. We encourage the participating teachers to become acquainted with several of the most interesting recent developments, so that they may use them as resources, of both ideas and materials, for their own creative work. Each of those new physics courses is in fact the result of a careful design effort to fit a given set of circumstances, and we try to encourage the setting up of local teams which will design their own courses. We wish to avoid generating two equally bad extremes: the "total adopter" and the "total ignorer." The first takes a foreign curriculum and adopts it in toto, uncritically, as it is. The second ignores all work done by others and thinks he can do it all by himself. The wise path, no doubt, lies between the two. And we hope that by letting a group of teachers work with several very different courses and discuss their relative merits we may increase the probability of creative work based on a critical analysis of the local needs and of the available pool of ideas and materials.

There are also other types of seminars and courses:

For instance, next year we shall be holding a rather special type of seminar at which we expect to have about 15 physics teachers working on the construction of problems and questions in physics and discussing the related objectives, teaching styles, student activities, and so forth. The purpose is twofold: partly to generate good problems and questions and have them used, but mainly to see how this activity can be a means for teacher training and for helping the teacher to open the way to curriculum reform. We owe this idea to Professor Eric Rogers, who will in fact be directing this seminar.

Another type of course we are planning for the near future is a special kind of internship program or postgraduate course for young scientists or science teachers who could eventually become the designers and the implementers of new science education systems. We think there is an increasing need for this kind of postgraduate program, especially for people in the developing countries, and we are planning to make suitable arrangements with institutions that excel in the field of science education, so that each intern may follow the course of studies and participate in activities that cater best to his needs.

We have singled out some subjects on which we would like to
carry out special studies. One of them, relevant to the present
discussion, will be the local production of simple low-cost apparatus, a problem of grave concern to developing countries. We
are increasing our activity in the field of integrated science, and
physics teachers interested in this approach will be welcome to
participate. UNESCO is also promoting the interest in extracurricular science activities, such as science clubs and science
fairs, and we shall try to get physics teachers involved in such
activities as well.

UNESCO is also assisting its member states with funds from
the United Nations Development Program in a variety of institution-
building projects of which I will mention those related to physics
teaching. For each of these, the United Nations provides funds
for consultants, fellowships, and equipment. In a typical project
this assistance lasts four or five years, and it is then carried on
as a purely local effort. Twenty-six teacher-training colleges
have been set up and are assisted in this way, 16 of them in Africa.
They all include also the training of teachers of science (hence also
physics). There are four university science faculties of institutes
that are being strenghtened in this way, as well as several institutes of technology; and two national Centers for Science Curriculum Improvement are also being supported.

International Atomic Energy Agency (IAEA)

The International Atomic Energy Agency (IAEA) and UNESCO have
been collaborating during the past two years on a program aimed
at identifying topics in nuclear science that could be included in
the science courses at secondary and university levels, and writing guidelines and planning the corresponding laboratory work. In
July 1968 a joint panel of consultants met in Bangkok (Thailand)
and devoted its attention to nuclear science as related to chemistry. A report was issued,[7] and a brochure on laboratory experiments was published subsequently.[8]

----

7. "Nuclear Science Teaching," Report of a Panel on Nuclear Science Teaching jointly convened by IAEA and UNESCO and held in
Bangkok, 15-23 July 1968. IAEA Technical Report No. 94, 1968.
8. "Experiments on Nuclear Science," UNESCO Chemistry Teaching Project in Asia, Bangkok, 1969.

As a sequence to the foregoing, these two U.N. agencies will hold a second panel of this kind in October 1970. This meeting will be discussing ways of introducing nuclear-science topics into general physics courses (late secondary and early university), and it is hoped that it will also issue guidelines on how to put their recommendations into effect. The participants at this meeting will be nuclear physicists with a strong interest and with personal experience in the improvement of science education.

Both IAEA and UNESCO collaborate with the University of Uppsala (Sweden) in the holding of a one-year postgraduate course in physics which takes place every two years at Uppsala.

World Meteorological Organization

The World Meteorological Organization (WMO) has recently brought out a publication under the title "Guidelines for the Education and Training of Meteorological Personnel,"[9] which outlines in great detail a variety of curricula (including laboratory work) for the training of various levels of meteorological personnel, both general and specialized. Pure and applied physics play a very important role in these education and training programs at the three stages covered: education in the basic sciences, fundamental meteorological education, and specialization.

Although these curricula are meant for training meteorological personnel, they could be of interest to some physics teachers who may wish to participate in such programs; and they are certainly useful to many of those responsible for the education of physics teachers. After all, there is a great need for more awareness by everybody of the deterioration that modern conditions are causing in our physical environment, mainly to our air and to our water. I am certainly not suggesting that this will be solved purely by knowing more atmospheric physics. We all know that we are facing here a variety of very complex and interrelated economic and social problems. But one of the many steps to be taken might well be the introduction into school curricula—whether in physics, in geography, or somewhere else—of relevant topics related to our atmosphere and our hydrosphere, at the appropriate level and through adequate examples and student activities.

----

9. WMO publication no. 258.TP.144, Geneva, 1969.

I would like to give you now two examples of what is being
done by regional organizations. One will be the Organization of
American States, which includes most of Latin America as well
as the U.S.A., and the other will be the Council of Ministers of
Education of Southeast Asia, which covers six countries in that
region.

Organization of American States

The Organization of American States has, among a great many
other activities, a series of programs in science and in science
education of which there are of course several of direct interest
to physics teachers. In recent years these programs have included such items as fellowships for physicists and physics teachers to spend a year or two in another country either within or outside the region; fellowships for physics teachers to attend summer
courses of 4 to 12 weeks' duration; visiting professorships for 5
to 12 months within the region; a series of special postgraduate
courses both for university lecturers and secondary-school physics teachers; and the Inter-American Conference on the Teaching
of Physics (held in Rio de Janeiro in June 1963 in collaboration
with the Latin-American Physics Center).

Apart from the foregoing, there are also technical-assistance
programs that cooperate with Ministries of Education on science-
curriculum projects and on the upgrading of science teachers. The
Organization is furthermore publishing a series of monographs
aimed at helping those who are teaching each of the sciences and
within which 11 titles in physics have been or are about to be published.

Council of Ministers of Education of Southeast Asia

The Council of Ministers of Education of Southeast Asia sponsors
a series of regional centers; one of these is the Regional Center
for Education in Science and Mathematics at Penang, Malaysia.
It serves six countries, and of course part of its activities are in
physics. The Center is in its initial stages of development and
has so far organized special refresher courses for teachers as
well as workshops on curriculum improvement. It is planning to
carry out a variety of activities aimed at improving education in

science and mathematics at the elementary and secondary levels and will collaborate in this with university faculties of science and with colleges of education. It will attempt to reappraise and modernize the subject-matter content, as well as to improve the technique for transmitting this to the pupil. For this purpose it will work with teachers on the proper design and use of instructional materials and on the use of the inquiry method and its relation to the learning process (including how this process varies among students). Apart from its training, research, development, and evaluation functions, the Center will also run information services for the benefit of teachers of the region in general and of leaders in science-curriculum improvement in particular.

Latin-American Physics Center

Let me now refer to a regional organization that is devoted specifically to the field of physics: the Latin-American Physics Center (Centro Latino-Americano de Fisica, CLAF). It was founded in 1962 as a result of steps initiated jointly by the Brazilian government and UNESCO. Its activities are of interest both to physicists and to teachers of physics. The Center has a fellowship program (134 fellowships during its first 8 years) and has so far sponsored 59 visiting professorships; it has cosponsored several important events in physics in Latin America, such as the yearly Escuela Latino-Americana de Fisica, the 1963 Inter-American Conference on the Teaching of Physics (with the Organization of American States), and the 1963 International Conference on Physics as Part of General Education (with IUPAP and UNESCO). The Center also participated with 14 fellowships in UNESCO's pilot project on the teaching of physics and one of its subsequent regional seminars. It publishes a periodical bulletin of interest both to research workers and to teachers, and is now also planning to carry out a regional survey of physics curricula and textbooks, with the aim of making the results available to those who are working in the region on the improvement of physics education.

East-African Secondary Science Project

One more example is a regional cooperative program covering three countries in Africa, which has the very specific aim of im-

proving science education at the secondary level. I am thinking
of the East-African Secondary Science Project. Part of its work
again refers to physics. The Ministries of Education of Kenya,
Tanzania, and Uganda have set up panels to supervise the work,
and these have been strongly supported by the science teachers'
associations, the university colleges, and the teachers' colleges
of these three countries. This regional project is rethinking the
content of the courses, the teaching methods, and the examinations; it is engaged in the writing of teachers' guides and textbooks and in the construction of visual aids and apparatus. It is
also working on the retraining of teachers through in-service
courses and is injecting new ideas into the teacher training colleges. The local production of physics-teaching apparatus is making good progress through the use of workshops in a university
college, a teachers' college, and a boys' school, as well as
through private firms.

The new physics course being developed by SSP, a four-year
course, is being tested in all three countries. The trials involve
110 teachers and 12,000 pupils in 64 schools. And the materials
are already being revised in the light of the feedback coming from
these experiments.

It is time now to ask some questions and to make some comments.

Let me first come back to the work that the International
Commission on Physics Education and UNESCO are doing for the
exchange of information through publications and meetings. Regarding the series "New Trends in the Teaching of Physics," for
instance, is it having a wide enough distribution? Is it available
in sufficient languages? Is it cheap enough? Does it come out fast?
Maybe there is also the need for another kind of publication, a
periodic newsletter, which could be produced fast and cheaply,
and which could be sent free of charge to a larger number of individuals and institutions? This suggestion has been made by several people in recent years. I think it deserves serious consideration.[10] Whether this newsletter is to be devoted to all the sci-

----

10. The "Working Party on the Improvement of Science Education
with Special Reference to Developing Countries," which was held
in Paris in September 1969 under the auspices of UNACAST and

ences or whether there should be separate issues for each of them is a question that requires further study.

I have said earlier in this paper that the "Survey of the Teaching of Physics in Universities" has been very successful and is still on demand. However, some years have gone by, and things are changing. If a follow-up to it were considered, it might well take the shape of a more critical survey and cover this time a number of university departments of physics in developing countries. I would try to make it a critical survey in the sense that it would attempt to show why some things went well and some went wrong. Such a survey might be very useful to those trying to set up physics departments in developing countries.

International meetings are a most important form of communication, especially if, as in the case of the present meeting, we can look forward to a report on it of equal clarity and relevance as those of the earlier meetings sponsored by the IUPAP Commission on Physics Education. The main worry at UNESCO in this connection continues to be the limited participation of physicists and physics teachers from the developing countries. In the meantime let us at least hope that the report in its final book form may have a wide circulation all over the world, including the developing countries. International seminars and special courses of interest to physics teachers are also being sponsored and will continue to be sponsored by international organizations; but there is of course room for more, especially for the benefit of teachers from developing countries.

International organizations are helping to promote and support innovative efforts in science education. I have given several examples of both small and large projects. Of course much more needs to be done; but I will add here only that perhaps not enough work and money are being put into the vital first step of identifying the potential local leaders and supporting their first efforts. Maybe a source of small grants that can be awarded quickly and flexibly is required. In dealing with such a variety of different situations all over the world, with different needs and different objectives, great care has to be exercised to avoid rigid schemes

----

UNESCO, also recommended the publication of such a newsletter as well as the continuation of the "New Trends" and "Sourcebooks."

leading to simple transfer solutions. Such mistakes have been
made, and they are as bad as trying to apply a physical law beyond its range of validity.

Another question relates to the professional associations of
science teachers. In some of the developing countries, such associations are already fairly active in promoting and applying innovations in science education. Perhaps international organizations could give some support to those science teachers' associations which need it and could stimulate the formation of such professional associations where they do not yet exist. This might
also give rise to the setting up of teachers' centers where the
teachers themselves would be responsible for their own improvement.

I have used the word "innovation" several times because
there seems to be no doubt as to the need for innovations. But
in many cases it is not even clear in which direction to innovate.
Much research is still needed in a variety of fields related to the
learning of physics and to learning in general; this research will
require the participation of scientists, teachers, psychologists,
media specialists, and others. To give just a few examples: for
the acquisition of a certain concept, which are the best combinations of manual, pictorial, verbal, mathematical, and other processes, and in what sequence can they be utilized most successfully? What are the optimum age and other circumstances at which
a certain concept can be used for the first time in a certain way?
What are the relations between personality factors of a pupil and
the kinds of instructional materials that are most suitable for him?
What are the best ways of utilizing individualized self-pacing
learning systems, and what will be the roles of the teaching personnel within them? Such research programs, and the comprehensive design efforts that will be needed for the development,
production, and evaluation of learning systems, could in many
cases benefit from international cooperation. This is particularly true in the case of a group of countries in a region that
shares similar problems. International organizations may therefore assist not only by stimulating such research and design work
but also by setting up international teams and in providing for the
exchange of information and discussion of results.

In several countries there already exist special institutions
that serve the purpose of demonstrating the utilization of modern

teaching methods and modern learning materials as well as performing special training and information tasks on a national scale.
International organizations in general, and UNESCO in particular,
could help to "internationalize" the work of these national centers
by promoting contacts between them and by supporting exchanges
that would get science education innovators from developing countries to come to these centers under special training programs.
The centers, although independent from each other, might in
some way constitute a kind of a network of institutions that could
be visited by anybody who wants to have firsthand knowledge on
the most interesting contemporary developments in science education.

I have the impression that not enough attention is being paid
in the world in general to the question of how to produce high-
quality science teachers. For instance, it seems to me that
teachers would be better prepared for teaching if during their
initial training at their colleges or universities to become teachers they were given opportunities for participating in the design
of new courses, in the preparation and testing of new learning
materials, in research on problems relevant to their future work,
and so on; and also if they were taught in the style in which they
are expected to teach. It seems to me that if we want the teachers to elicit creative behavior from their pupils, then the teachers themselves should also have opportunities to develop problem-
solving attitudes and to work creatively during their own training.

Let me now comment on the formation of leading innovators
in science education in the developing countries. How can they be
identified and helped? What should they do as part of their education and training? Where will they do it? How do we make sure
that what they learn is relevant to their countries and will help
them face their own problems? And this is only a particular case
of a more general one: that of the young graduates who are sent
to pursue postgraduate studies in any of the more advanced countries and are put to work on something of interest to their hosts
but not necessarily to themselves. At that very moment they become potential candidates for what is know as the brain drain.
The organizations that provide the fellowships and the institutions
that receive the fellows will have to pay more attention to this;
and they may even wish to experiment with more flexible admin-
istrative arrangements that may permit, for instance, fellowships

including alternating periods of study abroad and work at home.
This may not be easy to arrange, but it is worth trying.

I have tried to indicate what kinds of efforts are being made
and have briefly touched on some of the most pressing questions.
What seems so striking to me is that all this work is so very recent; it has hardly begun to scratch the surface. There is so much
more that has to be done, but it is not just a question of doing
"more of the same thing." Fortunately there is now an increasing
awareness of the need to improve the quality of education, not
just its quantity. This will require much experimentation with
new approaches and new styles in teaching and learning. International organizations have a very important role to play in this—
in fact a double role: on the one hand to stimulate a diversity, of
approaches and the confrontation of results, and on the other hand
to coordinate so as to avoid dispersion of efforts and wasteful
repetition. In so doing, international organizations will also continue to try to help people all over the world to understand each
other better.

List of Contributed Papers and Documents

A. V. Baez: "Film Loops in Physics Produced by Encyclopaedia Britannica Educational Corporation in Collaboration with Dr. A. V. Baez."

W. A. Blanpied and V. S. Nigama: "A Note on the Indo-U.S. Summer Institute Programme for Secondary School Physics Teachers."

P. E. Blosser and R. W. Howe: "An Analysis of Research Related to the Education of Secondary School Science Teachers."

S. C. Brown and N. Clarke: International Education in Physics (Proceedings of the Paris Conference) M.I.T. Press, 1960. Resolution on pages 1-3, and Chapter 7, pages 73-87.

S. C. Brown and J. Zemplén: "Report on the International Working Seminar on the Role of Physics in Physics Education, Cambridge, Massachusetts, July 1970."

J. Casanova: "The Spanish School of Secondary Physics Teachers."

J. Cessac: "La formation pédagogique des professeurs de science physique des lycées."

J. Cessac: "Les centres d'équipement en matériel scientifique."

B. R. Chapman: "Towards a Satisfactory Model of the Education of Physics Teachers."

G. Cortini and G. Segre: "In-Service Education of Teachers at the University of Naples."

J. B. Cross: "A New Four-Year University Curriculum for Future Teachers of Physics and Chemistry in Secondary Schools."

R. David: "La préparation d'un montage ou d'une leçon type
CAPES en agrégation donne-telle actuellement une formation
professionelle suffisante à un professeur de sciences physiques
dans l'enseignement secondaire?"

Droujba Conference 1968: Integration of Science Teaching, General Report, pages 9-37.

O. Eisler: "On 'Vertical Links' Between School Physics and Advanced Physics."

L. R. B. Elton, P. J. Hills, and S. O'Connell: "Self-Teaching
Situations in a University Physics Course for Secondary School
Physics Teachers."

W. Eppenstein: "Summer Institutes for Secondary School Physics Teachers."

P. Fleury: "Sur la préparation des maîtres à leur rôle d'examinateurs."

P. Fleury: "Les présentations d'experiences dans l'enseignement
de la physique au niveau secondaire."

F. H. Giles: "The Training, Tenure, Schedules and Opportunities of and for Physics Teachers in the Secondary Schools of a
Typical Southeastern State in the United States."

G. G. Gluck: "La préparation du jeune professeur à la rédaction
d'exercises."

A. Harashima: "Use of computer-generated animation films in
the teaching of physics in secondary schools."

K. Hinst: "Educational Technology, Implications for the Educational System, National Policies, and the Role of the Teacher in
an Individualized Learning System."

The Institute of Physics, The Physical Society, and The Royal
Society, London, 1969: "Teacher Training for Physics Graduates:
A Commentary."

N. Joel: "How Can International Organization Help Physics Teachers?"

W. V. Johnson: "The Role of the Regional Physics Associations in the United States."

Y. Kakiuchi: "College Physics for the Future School Teachers: its Implication and Approach."

P. L. Kapitza: "General Principles of the Education of Present-Day Youth and General Methods of Secondary-School Physics Teaching."

R. L. Krans: "Semantic Difficulties for the Young Pupil 13 to 15 Years in the Beginning of Learning Physics."

R. L. Krans: "School Physics as a University Specialism."

W. Kroebel: "The Training of Physics Teachers for Secondary Schools and the Dependence of this Training on the Instruction in Universities."

E. B. Kvashin: "The Training of Qualified Physics Teachers in Secondary Schools in Nigeria."

L. Leboutet: "Réactions des adolescents à l'enseignement de la physique."

P. G. de Paula Leite: "Observations on the Teaching of Physics in Developing Countries." Why Teach Physics? (Proceedings of the Rio Conference) S. C. Brown, N. Clarke, and J. Tiomno; M.I.T. Press 1964, pp. 11-12.

R. N. Little: "Pre-Service Formation of Physics Teachers: Technical Education."

A. Loria: "In-Service Education of Physics Teachers at Modena University."

J. T. Macfarlane: "The Special Problems of the Secondary School Physics Teacher in Developing Countries."

G. Marx: "An Insoluble Task: Teaching Physics."

M. R. Mayfield: "Physics: The Program for Teachers. Pre-Service Preparation of High School Teachers."

H. Messel: "A Look at the New Integrated and Coordinated Science Courses in N.S.W."

R. J. Miller: "The Illinois State Physics Project. The Involvement of Illinois Colleges and Universities in Cooperatively Improving Secondary Physics Education."

S. O'Connell, A. W. Wilson, and L. R. B. Elton: "Preknowledge Survey for University Science Entrants." Nature 222, 526 (1969).

Organization for European Cultural Development (O.E.C.D.): "Educational Technology—A Systematic Approach to the Teaching/Learning Process" (1969).

D. Pescetti: "Development of Mass Instruction Techniques; Use of the Computer as Teaching Aid."

A. D. Pickar: "A College Course in Physics, Chemistry and Biology as a Preparation for Secondary School Teachers."

E. M. Rogers: "The Nuffield Project." Physics Today, 20, 40 (1967).

E. M. Rogers: "The Use of New Examinations and of Examination-Construction Seminars in Curriculum Revision and in the Training of Teachers."

The Royal Society, London, 1969: "The Shortage of Mathematics and Science Teachers in Schools."

The Royal Society, London, 1970: "Teacher Training for Science and Mathematics Graduates."

The Royal Society, London, 1970: "In-Service Training for Teachers of Mathematics and Science in Schools."

H. Sauvenier: "Tendances nouvelles dans la formation des professeurs de physique en Belgique."

S. Sikjaer: "The Work and Duties of the Physics Institute of the Royal Danish School of Educational Studies."

M. W. P. Strandberg: "Technology in Education."

D. A. Tawney: "The Role of Physics Centers in the Initial and In-Service Training."

M. Underwood: "On-Going Curriculum Development and the Problem of Achieving it in the United Kingdom."

R. C. Waddell: "The Program of Involvement of Eastern Illinois University (U.S.A.) as an Example of the State Program."

F. Watson: "Pre-Service Pedagogical Formation of Physics Teachers."

E. J. Wenham: "The In-Service Education of Physics Teachers."

E. A. Wood: "Pressing Needs in School Science." American Institute of Physics, 1969.

P. Youngner: "A Different Sequence for Science Courses in the High Schools of the United States."

Names and Addresses of Participants

Belgium
M. Lemaitre
Tervuurse Vest 123/12
3030 Heverlee

L. A. Verhaegen
Afdeling Didaktische Fisica,
Departement Natuurkunde,
Katholieke Universiteit
Naamsestraat 61
3000 Leuven

Brazil
R. B. da Costa
Centro Latino Americano
de Física
Av. Wenceslau Bráz 71-ZC 82
Rio de Janeiro

E. W. Hamburger
Instituto de Fisica,
Universidade de São Paulo
Caixa Postal 8105
São Paulo

Bulgaria
V. Metev
Research Institute
of Education
Boul. Lenine N 125, Blok 5.
Sofia

P. Targov
Boul. "Moskva" 172
Plovdiv

T. Vassilev
R'Université de Plovdiv
Ul. Zar Assene 24.
Plovdiv

V. Weltchev
Ministry for Public Education
Sofia

Czechoslovakia
Mrs. M. Chytilovà
Mikulandska 5
Praha 1

K. Dudás
Samorin

J. Fuka
Universita Palacky
Leninova 26
Olomuc

Mrs. J. Hnilickovà
Prikra 16/271
Praha 4 - Branik

J. Vachek
Ke Karlovu 3
Praha 2

Denmark
O. Eisler
Institute of Physics,
Aarhus University
Aarhus

S. Sikjaer
Harsdorffsvej 6A
1874 Copenhagen V

P. Thomsen
Kornagervej 10
2800 Lyngby

Federal Republic of Germany
H. Brockmeyer
Hahnredsweg 14
3562 Wallau (Lahn)

W. Kroebel
Institut fur Angewandte Physik
Neue Universität
Ohlshausen Str.
23 Kiel

S. Sotier
Robert Koch Str. 5
8 München 22

H. Volz
Winkelweg 7
852 Erlangen

France
J. C. Beaufils
Faculté des Sciences
Physique Fondamentale
P 5, B.P. 36
59 Lille

M. Y. Bernard
Department of Electronics
Conservatoire National
des Arts et Métiers
292 Rue Saint-Martin
75 Paris 3e

F. Blain
138 Avenue du
Général Leclerc
91 Gif sur Yvette

J. Boutigny
23 Rue du Bois Robert
78 Saint-Cyr l'École

J. Cessac
115 Avenue de Paris
78 Versailles

Mrs. M. Cordier
5 Rue Boucicant
92 Fontenay-aux-Roses

R. David
63 Avenue Victor-Hugo
92 Claramart

P. Fleury
Institut d'Optique
3 Boulevard Pasteur
75 Paris 15e

G. G. Gluck
Physique Agrégation
Faculté des Sciences
Caen 14

Mrs. L. Leboutet
16 Résidence Beausoleil
92 St. Cloud

P. Martinot-Lagarde
Service Agrégation Faculté
des Sciences d'Orsay
91 Orsay

R. Petit
Faculté de Saint-Jerome
13 Marseilles 13^e

German Democratic Republic
H. Hänsel
Pädagogische Hochschule
Potsdam,
Sektion Mathematik-Physik
15 Potsdam

K. Jupe
Universität Jena,
Sektion Mathematik
August-Bebel Str. 4
69 Jena

H. Melcher
Wattstr. 5
1502 Potsdam-Babelsberg

Hungary
Mrs. I. Abonyi
Váci u. 85
Budapest V.

G. Antal
Bolyai Gimnázium
Salgótarján

I. Bayer
National Pedagogical Institute
Gorkij fasor 17-21
Budapest VII.

I. Berzi
Thomas Mann u. 9
Dabrecen

I. Bukovszky
Budafoki ut 8
Budapest XI.

B. Deczky
Dózsa György ut 23
Debrecen

Z. Demendy
Nehézipari Müszaki Egyetem
Fizika Tanszék
Miskolc-Egyetemváros

I. Duchnovszky
Tanárképző Főiskola
Fizika Tanszék
Ifjuság u. 6
Pécs

B. Fejér
Széchenyi u. 19
Eger

J. Firtkó
Nehézipari Müszaki Egyetem
Fizika Tanszék
Miskolc-Egyetemváros

B. Fogarassy
Institute for Experimental
Physics
Eötvös University
Muzeum krt. 6-8
Budapest VIII.

I. Főzi
Institute for Experimental
Physics
Eötvös University
Muzeum krt. 6-8
Budapest VIII.

A. Frank
Perczel Mór Gimnázium
Siófok

L. Holics
Dávid Ferenc u. 7
Budapest XI.

Mrs. S. Kántor
Fazekas Gimnázium
Hatvan u. 44
Debrecen

F. J. Kedves
Institute for Applied Physics
L. Kossuth University
Debrecen 10

F. Kiss
Árpád sor 84/1
Békéscsaba VI.

Miss Gy. Kiss
Halmi u. 29
Budapest XI.

Z. Kiss
Május 1 ut 12
Kaposvár

L. Kovács
Landler Jenő Gimnázium
Vöröshadsereg u. 9
Nagykanizsa

B. Lechoczky
Marx ut 34
Dombóvár

Mrs. Gy. Leitner
Beloiannisz u. 5
Eger

P. Mag
Department of Atomic Physics
Eötvös University
Puskin u. 5-7
Budapest VIII.

G. Marx
Department of Atomic Physics
Eötvös University
Puskin u. 5-7
Budapest VIII.

T. Mátrai
Pedagogical Institute
Eger

E. Nagy
Institute for Experimental
Physics
Eötvös University
Muzeum krt. 6-8
Budapest VIII.

P. Nagy
Táncsics Gimnázium
Orosháza

Á. Nyitrai
Váci ut 21
Budapest XIII.

Gy. Patkó
Pedagogical Institute
Eger

I. Pahán
Rákóczi u. 7
Nagykörös IV.

Mrs. Gy. Pálffy
Tanárképző Főiskola
Pécs

T. Pánczél
Ady Endre u. 41
Fonyód

L. Párkányi
Institute for Experimental
Physics
Eötvös University
Muzeum krt. 6-8
Budapest VIII.

I. Poór
Institute for Experimental
Physics
Eötvös University
Muzeum krt. 6-8
Budapest VIII.

E. Sas
Institute for Experimental
Physics
Eötvös University
Muzeum krt. 6-8
Budapest VIII.

L. Skrapits
Institute for Experimental
Physics
Eötvös University
Muzeum krt. 6-8
Budapest VIII.

A. Szabó
Gimnázium
Derecske

Miss P. Szabó
Budapesti Müszaki Egyetem
Kisérleti Fizika Tanszék
Budafoki ut 8
Budapest XI.

D. Szilágyi
Kossuth u. 67
Debrecen

M. Szombothy
G. Gárdonyi Gimnázium
Eger

P. Tasnádi
Institute for Experimental
Physics
Eötvös University
Muzeum krt. 6-8
Budapest VIII.

F. Tihanyi
National Pedagogical Institute
Gorkij fasor 17-21
Budapest VII.

Mrs. S. Tóth-Pál
Institute for Experimental
Physics
Eötvös University
Muzeum krt. 6-8
Budapest VIII.

L. Varga
National Pedagogical Institute
Gorkij fasor 17-21
Budapest VII.

L. Varga
Tanácsköztársaság utja 62
Debrecen

L. Vize
Institute for Experimental
Physics
A. József University
Dóm tér 9
Szeged

Mrs. L. Vize
Institute for Experimental
Physics
A. József University
Dóm tér 9
Szeged

L. Wiedemann
Institute for In-Service
Education of Teachers
Horváth Mihály tér
Budapest VIII.

J. Zemplén (Mrs. L. Mátrai)
Budapesti Müszaki Egyetem
Kisérleti Fizika Tanszék
Budafoki ut 8
Budapest XI.

India
W. A. Blanpied
National Science Foundation
9 Ring Road
Lajpat Nagar IV.
New Delhi 24

Ireland
S. O'Donnabháin
Department of Education,
Secondary Branch
Hawkins House
Dublin 2

Italy
G. Cortini
Istituto di Fisica Sperimentale
Via A. Tari 3
80138 Napoli

Mrs. M. Ferretti
Via Luca Ghini 6
40136 Bologna

A. Loria
Istituto di Fisica
Via Università 4
41100 Modena

Mrs. M. B. Palma-Vittorelli
Istituto di Fisica
Via Archirafi 36
Palermo

D. Pescetti
Istituto di Fisica
del' Università
Viale Benevetto XV, 5
16132 Genova

D. Sette
Istituto di Fisica
Facoltà de Ingegneria
Città Universitarià
Piazzale delle Scienze 5
00100 Rome

Japan
A. Harashima
International Christian
University
3-10-1, Osawa, Mitaka
Tokyo

Y. Kakiuchi
Institute for Solid State
Physics
University of Tokyo
22-1. 7-Chome, Roppongi
Minato-Ku, Tokyo

Malagasy Republic
B. Dolphin
B. P. 1500
Tananarive

Netherlands
F. Balkema
Van der Helstlaan 20
Naarden

H. P. Hooymayers
Didactic Department of
Physics Laboratory
State University of Utrecht
Transitorium I
Leuvenlaan
Utrecht

J. S. Kobus
Malmberg Educational Publ.
16 Leeghwaterlaan
S'Hertogenbosch

R. L. Krans
Didactic Department of
Physics Laboratory
State University of Utrecht
Transitorium I
Leuvenlaan
Utrecht

J. B. Van der Kooi
De Esstukken 8
Haren (GR)

Nigeria
J. A. O. Sofolahan
Ministry of Education
Ibadan

Poland
Mrs. M. Bochenek
Al. Niepodlegtosci
Nr. 222 m. 15
Warszawa 1

Mrs. D. Stachorská
Uniwersytet Maria Curie-
Slodowska
Instytut Fizyki
Ul. Nowotki 8
Lublin

Rwanda
J. T. Macfarlane
Université Nationale du Rwanda
B. P. 117
Butare

Spain
J. Casanova
Facultad de Ciencias
Valladolid

E. G. Nagore
Colón No. 7
Valencia

Sweden
G. Brogren
Chalmers University
of Technology
Gothenburg

Switzerland
H. H. Staub
Physik-Institut der
Universität Zürich
Schönberggasse 9
8001 Zürich

Union of Soviet Socialist
Republics
S. G. Bronevshuk
Ministry of
Education U.S.S.R.
Chistyie Prudy 6
Moscow

P. L. Kapitza
S. I. Vavilov Institute of
Physical Problems
Vorobievskoie Shosse 2
Moscow V-334

Mrs. E. A. Shekuteva
Voroshilovsky Raion
School No. 154
Moscow D-98

Mrs. G. S. Tarasjuk
Department of Physics
Lomonosov University
Moscow V-234

United Arab Republic
Mrs. I. S. Tawfik
Senior Inspector of Science
for Ministry of Education
Mogamma El Tahrir
Cairo, Egypt

A. E. M. El Kashef
West Educational Zone of Cairo
7th floor of Mogamma El Tahrir
Cairo, Egypt

United Kingdom
B. R. Chapman
Centre for Studies in Science
Education
The University
Leeds 2

E. Eisner
Department of Applied Physics
University of Strathclyde
Glasgow C 1

L. R. B. Elton
Institute for Educational
Technology
University of Surrey
Guildford
Surrey

P. L. Flowerday
The Institute of Physics
47 Belgrave Square
London

D. W. Harlow
The Royal Society
6 Carlton House Terrace
London SW 1.

H. F. McMahon
Educational Centre
The New University of Ulster
Coleraine, Northern Ireland

E. M. Rogers
Physics Department
Princeton University
Princeton, N.J. 08540,
U.S.A.

H. Silver
Centre for Science Education
Chelsea College
Bridges Place
London SW 7.

D. A. Tawney
University of Keele
Keele
Staffordshire ST5 5Bb.

M. Underwood
University of London,
Institute of Education
Malet Street
London WS 1.

E. J. Wenham
Worcester College
of Education
Henwick Grove
Worcester WR2 6AJ

United States of America
A. V. Baez
La Rancheria
Carmel Valley, Cal. 93924

W. A. Blanpied
See under India

S. C. Brown
Department of Physics
Massachusetts Institute of
Technology
Cambrige, Mass. 02139

S. C. Chen
Department of Physics
Drexel University
Philadelphia, Pa. 19104

J. B. Cross
Education Development Center
55 Chapel St.
Newton, Mass. 02160

W. Eppenstein
Department of Physics
Rensselaer Polytechnic
Institute
Troy, N.Y. 12181

H. C. Jensen
Lake Forest College
Lake Forest, Ill. 60045

W. V. Johnson
Americal Association of
Physics Teachers
1785 Massachusetts Ave., N.W.
Washington, D.C.

W. C. Kelly
National Research Council
2101 Constitution Ave.
Washington, D.C. 20418

R. N. Little
Physics Department
The University of Texas at Austin
Austin, Texas 78712

M. R. Mayfield
Department of Physics
Austin Peay State University
Clarksville, Tenn. 37040

R. J. Miller
Greenville College
Greenville, Ill. 62246

O. Oines
Glenbrook North High School
Northbrook, Ill.

L. W. Phillips
Division of Undergraduate
Education in Science
National Science Foundation
Washington, D.C. 20550

A. D. Pickar
Department of Physics
Portland State University
Portland, Oregon 97201

T. Porter
Division of Graduate
Education in Science
National Science Foundation
Washington, D.C. 20550

O. P. Puri
Cooperative General Science
Project
Clark College
Atlanta, Ga. 30314

M. W. P. Strandberg
Department of Physics
Massachusetts Institute of
Technology
Cambridge, Mass. 02139

R. S. Tilton
Chairman of Science Dept.
School Department
Cicero, N.Y. 13212

R. C. Waddell
Eastern Illinois University
Charleston, Ill. 61920

G. Walker
Department of Physics
Clark College
Atlanta, Ga. 20314

F. Watson
Harvard Graduate School of
Education
Longfellow Hall
Cambridge, Mass. 02138

P. Youngner
Department of Physics
State College
St. Cloud, Minn. 56301

Venezuela
J. A. Rodriguez
Departamento de Matematica
y Fisica
Instituto Pedagogico
Avenida Paez. el Paraiso
Caracas

Yugoslavia
E. Danilović
Jugoslovenski Zavod za
Proucavanse Skolskih i
Prosvetnih Pitanja
Draze Pavlovica 15
Beograd

International Organizations
K. Hinst
Center for Educational Re-
search and Innovation
Organization for European
Cultural Development
(O. E. C. D.)
2 Rue André Pascal
75 Paris 16e, France

N. Joel
Division of Science Teaching
UNESCO
Place Fontenoy
75 Paris 7e, France

Appendix C

Curriculum Projects

The following material was taken mainly from articles in Phys-- ics Today, March 1967, Vol. 20, in which PSSC is described by Uri Haber-Schaim, the Nuffield Project by Eric Rogers, and Harvard Project Physics by Gerald Holton.

PSSC

The Physical Science Study Committee was a group formed in the United States in 1957 to design an entirely new physics course for secondary-school students in the eleventh or twelfth grade, 17 to 18 years old. The project was funded by the National Science Foundation (NSF). At the end of 10 years more than half of the high-school physics students in the U.S.A. were enrolled in PSSC courses. Eighteen foreign editions of the textbook had been published, and the laboratory guide was available in 15 languages. The training and retraining of teachers had been started through summer and evening programs financed by NSF, and the widely used College Entrance Examination Board achievement tests included an alternative specifically intended for students who had studied physics under this system.

The goals of the Committee determined not only major decisions about the course but also many details about apparatus, about which topics should be covered in laboratory work and in the classroom, and what material could best be presented in films. These were the goals:

1.
To present physics as a unified yet living and ever-changing subject.

2.
To demonstrate the interplay between experiment and theory in the development of physics.

3.
To have the students learn the basic principles and laws of physics by interrogating nature itself, thus learning not only the laws but also the evidence for them as well as their limitations.

4.
To extend the student's ability to read critically, to reason, and
to distinguish between the essential and the peripheral, thereby
improving his learning skills in general.
5.
To provide a sound foundation for those students who plan to study
science or engineering at the college level.

To demonstrate the unity of physics, PSSC concentrates on
principles that can be seen on the astronomical as well as on the
human and atomic scales. The course is based on two main subjects: the dynamics of particles under the influence of forces and
conservation laws of momentum and energy, and the superposition principle governing all wave propagation.

Thus Newton's second law of motion is studied in laboratory
experiments with carts, rubber bands, and timers, and then
applied to both planets and alpha particles. Conservation of momentum is investigated with ball bearings and later applied to
rockets and photon scattering by electrons. Interference of waves
is observed in the ripple tank and applied to light passage through
a double slit and to electron reflection from a crystal.

To demonstrate the interplay between experiment and theory,
the pupil studies in detail the evolution of theories of light. After
investigating the behavior of light in the laboratory, he considers
the particle theory of light, abandons it in favor of the wave model,
and finally returns to it in a modified form.

In order to have the student experience for himself in a valid
way the process by which concepts develop from practical work,
his experience in the laboratory must be positive rather than
simply descriptive. The teacher's guide offers flexibility by suggesting alternatives in the scheduling of particular experiments.
Since the whole class must work at the same time on an investigation to develop results together, apparatus was designed simple
and strong enough to be used repeatedly by students and inex-
pensive enough so that schools could purchase enough of each
piece of equipment to allow only two to four students to share.

Films show experiments beyond the reach of classroom teaching. Several physicists were invited to participate in the filmed
demonstrations, serving the double purpose of showing an experiment and also a variety of physicists themselves at work.

The PSSC course occupies five to seven one-hour periods of classroom and laboratory time per week, and assumes no significant preparation on physics. Students are expected to have studied one year of algebra and one year of geometry in secondary school

Further information can be obtained from conference member J. B. Cross of the Education Development Center, Newton, Massachusetts, U.S.A.

The Nuffield Project

Begun in England in 1962 under the sponsorship of the Nuffield Foundation, this project set out to design a new and more imaginative science curriculum for English schools, including physics, chemistry, and biology, to run through the five years from age 11 to ages 16 or 17. The program was planned by teachers for teachers, and a central objective was "to develop science for all in the grammar school group, not merely for the future specialist but for the future citizen in the latter half of the twentieth century."

The pupil's view of the subject was to be a main concern. Science should become intellectually exciting for him, and he should come through his own investigations and discussions to understand what science is, and as much as possible to feel what it would be like to be a practicing scientist. The major goal was to be "the achievement of an understanding deeper than the mere ability to repeat what the pupil has been carefully instructed to repeat."

In the physics part of this curriculum, the scheme runs through the five years in interwoven threads, building up knowledge of matter, waves, energy, and atoms. The main topics are:
1.
Forces and motion: from a simple study of springs in the first year to Newton's laws in the third and fourth years, which are used in the fifth year in planetary astronomy to show the development of theory.
2.
Electricity and magnetism: very simple circuits in the second year to voltmeters and electromagnetic induction in the fourth year.

3.

Waves and optics: ripple tank and optical instruments in the third
year to diffraction gratings in the fifth. Two other major strands
weave through these topics:

4.

Atoms and molecules: from crystals and models and measurement of oil molecules (each student for himself) in the first year,
to radioactivity and the Rutherford model in the fifth year, with
considerable attention to the kinetic theory of gases in between.

5.

Energy: from simple first acquaintence to a variety of sophisticated uses.

The first two years aim at getting the children acquainted
with the feel of science, mostly through the experiments conducted by the children themselves. They learn words such as
"atom" and "energy" by using them much as a small child learns
language by using it. Teaching is more formal after the second
year, when students use algebra and geometry freely and are
stimulated to more and more reasoning and to developing simple
examples of theory.

Apparatus was very carefully considered. Simple robust
forms were devised and manufacturers urged to produce quickly
and in quantity. Kits were boxed in sets of 16 (for a class of 32)
to make it difficult for teachers to convert a laboratory experiment into a demonstration. Many of the experiments were designed deliberately to let children discover how it feels to be a
scientist rather than to yield some particular measurement or to
verify some law.

Although the Nuffield Project course was planned explicitly
for the English school system and needs considerable adapting
to be used elsewhere, the teachers' guides offer suggestions
which would make them valuable in teachers' libraries in other
countries.

Further information can be obtained from conference member E. M. Rogers (U.K.) at his Princeton, N.J., address.

Harvard Project Physics

The planning for Harvard Project Physics began in 1964 and the
final version is now available. Like the Nuffield Project in

England, the developers were mainly concerned that this introductory course should serve the general student as much as the
future scientist, and that the result be not only some knowledge
of physics but an understanding of how science and society relate to each other.

The course is designed for one year in any school, but its
six units can be completed in six to eight months under the guidance of an experienced teacher or with an above-average class.
In these cases the teachers and class are free to choose from a
variety of supplemental topics, according to their own interests,
thus permitting more exploratory opportunities than has usually
been the case.

In the first unit the student is introduced to the concepts of
motion. The second unit studies celestial motion, following historically the development of the idea that the same principles of
motion apply to the planets as apply on earth. The third unit
proceeds to the conservation laws of momentum and mechanical
energy, the first law of thermodynamics and some discussion of
the second law. The fourth unit presents electricity and magnetism in the context of fields at rest and in motion, and traces the
subsequent failure of the mechanistic view. The origins of the
new physics and the atomic and nuclear models of matter are introduced in the last two units.

The course features from time to time the interdependence
of physics and the other sciences, and of science and other endeavors of human affairs more broadly. For example, in discussing the laws of thermodynamics the point is made, in the
student guide but more extensively in other readings, that the
heat engine was not just a technical achievement but a major influence in changing the structure of western society during the
Industrial Revolution and affected the imaginations of poets and
theologians as well as of mathematicians.

The student textbook in Project Physics is only one of the
materials provided, and a strong effort is made to ensure that
teachers and students make use of a wide group of other resources. These include a student handbook, laboratory equipment, programmed instruction, films, film loops and transparencies. There are 16 mm sound films to help in teacher training.
A teacher's guide for each unit discusses the use of the other materials such as the laboratory experiments, demonstrations and
films, as well as the history of science background. The course

seeks to make the student aware of the humanistic and cultural aspects of physics by avoiding overspecialized topics and by making use of history of science as a pedagogic aid whenever appropriate.

Further information can be obtained from conference member F. Watson of the Harvard Graduate School of Education, Cambridge, Massachusetts, U.S.A.

Teaching Films Presented

from A. V. Baez (U.S.A.):
Pulses and Waves
Superposition of Pulses in a Spring
Superposition of Pulses (Computer Animated Film)
Transverse Standing Waves in a Spring—Continuous Wave Trains
Ripple Tank and Soap Film—Standing Waves
Chladni Plates—Standing Wave Patterns
Conductors, Insulators, and Capacitors
Coulomb's Law
Discharging the Electroscope—Photoelectric Effect

from A. Harashima (Japan):
Computer (how animated films are made)
Motion of Projectiles (with no air resistance)
Motion of Projectiles (with air resistance)
Planetary Motion
Relativity I, II, and III
Classical Rutherford Scattering
Rutherford Scattering
Photons and Waves A and B
Electrons and Waves

from H. C. Jensen (U.S.A.):
Vector Addition; Velocity of a Boat
Conservation of Energy; Pole Vault

from D. Pescetti (Italy):
Magnetic Domains and Magnetization Processes in a Gadolinium
    Iron Garnet $(3Gd_2O_35Fe_2O_3)$

from F. Watson (U.S.A.):
Standing Electromagnetic Waves
Superposition

The Conference Exhibition

The Conference exhibition was arranged under the chairmanship
of Professor T. Mátrai and was open all during the Conference.
Coworkers of Professor Mátrai were Mrs. Hriczó, Mr. Patkó,
and Mr. Szombathy, joined by Professor H. C. Jensen (U.S.A.).
   The purpose of the exhibition was to bring a sampling of
modern teaching and learning aids to the attention of the participants. These aids included laboratory apparatus, demonstration
equipment, textbooks, laboratory manuals, workbooks, learning
programs, film and film loops and strips, overhead transparencies, and so on. Both individual and commercial equipment were
shown, and some of the highlights were the following:

H. C. Jensen (U.S.A.)

An apparatus to show mechanical vibrations in a copper spring.
Current passing through a helical wire is modulated by a low-
frequency generator. At the lower end of the wire is a permanent magnet, and in the field of this the wire moves periodically.
Adjusting the frequency appropriately, standing waves are observed along the axes of the helical wire.
   Jensen also demonstrated a recorder with which it was possible to record differences down to 10 millivolts with the output
recorded on paper tape.

A. Leitner (Department of Physics, R.P.I., Troy, N.Y., U.S.A.)

Equipment to show the vibrational modes of a rubber membrane.
A circular rubber sheet was fastened to a ring frame, diameter
30 centimeters; the vibrations are produced by a loudspeaker fed
by a tunable low-frequency generator.

H. Waage (Department of Physics, Princeton University. Princeton, N.J., U.S.A.)

Demonstration of the superposition of mechanical vibrations.
Mirrors were attached to two pendulums having variable vibrational frequencies. A light spot is reflected from both mirrors
onto a screen. The superposition is observed by means of a third
mirror that reflects part of the beams from the first mirror onto the second. The beam reflected by the second mirror shows
the superimposed vibrations.

Electromagnetic induction. A coil producing a magnetic field
is fed through a linear potentiometer driven by a reversible electric motor, so that the current passing through the coil increases
and decreases linearly with time. In a second coil one can measure the induced e.m.f. which is constant in the constant speed
region but changes rapidly at the ends of the potentiometer.

L. R. B. Elton (U.K.)
Exhibit of a series of pictures and slides showing modern teaching aids at the University of Surrey.

A. Harashima (Japan)
Exhibit of slides, transparencies, and film loops from the International Christian University in Tokyo.

Sargent-Welch Scientific Company (7300 North Linder Ave.
    Skokie, Illinois, U.S.A.)
Ripple tank. Waves are produced by a small electric motor. The
phase difference between two sources is variable. The exhibit
demonstration included the generation and interference of ring
waves, and the refraction and diffraction of plane waves.
    Also shown were an optical kit, a ribbon timer, a Millikan
apparatus, and equipment for measuring Planck's constant.
    This company sells a kit for nuclear physics produced by the
Union Carbide Corporation which allows one to make 25 different
experiments in the classical parts of nuclear physics.

Pasco Scientific (1933 Republic Ave., San Leandro, California,
    U.S.A.)
Electrostatic experiments. The basic element is a simple but
sensitive electrometer, adjustable by light spot or manually. It
can measure dc voltages from 10 mV to 100 V, and dc currents
from 10^{-13} to 10^{-4} amperes.
    A Millikan apparatus.

Damon Engineering Educational Division (115 Fourth Ave.,
    Needham Heights, Massachusetts, U.S.A.)
Kit for nuclear physics, as well as smaller teaching aids such
as a balance kit, a gas simulator having a variable-speed elec-
tric-drive motor, and a simple grating spectroscope.

Didafrance (16 Rue Saint-Denis, Paris, France.)
Air track. To demonstrate motion and collisions, small neon
flash lamps mounted on the sliders were fed by a frequency generator 50 or 100 Hz, and the flashing light spots were recorded
photographically. The equipment included a Land camera.

Also shown, a decade scaler with radioactive source and
an apparatus for demonstrating mechanical resonance.